Key Concepts in Graph Theory

Key Concepts in Graph Theory

Edited by **Jen Blackwood**

CLANRYE
INTERNATIONAL

New Jersey

Published by Clanrye International,
55 Van Reypen Street,
Jersey City, NJ 07306, USA
www.clanryeinternational.com

Key Concepts in Graph Theory
Edited by Jen Blackwood

International Standard Book Number: 978-1-63240-322-3 (Hardback)

Contents

Preface

This book introduces key concepts in graph theory, which is considered to be a crucial analytical tool in computer science and mathematics. Due to the built-in simplicity of graph theory, it can be employed to model several distinct physical and abstract systems like transportation and communication networks, models for business administration, psychology, and political science and so on. The aim of this book is not just to describe the present state and development tendencies of this theory, but to educate the reader enough to enable him/her to embark on the research complications of their own. Taking into consideration the huge amount of knowledge regarding graph theory and its practice, this book focuses on the applications of graph theory in future electric networks, power systems, algorithms and communication networks. This book intends to serve as a valuable source of reference for students associated with various fields like system sciences, engineering, social sciences, mathematics, computer sciences, etc. as well as for practitioners and software professionals.

Significant researches are present in this book. Intensive efforts have been employed by authors to make this book an outstanding discourse. This book contains the enlightening chapters which have been written on the basis of significant researches done by the experts.

Finally, I would also like to thank all the members involved in this book for being a team and meeting all the deadlines for the submission of their respective works. I would also like to thank my friends and family for being supportive in my efforts.

Editor

Power Restoration in Distribution Network Using MST Algorithms

T. D. Sudhakar
St Joseph's College of Engineering, Chennai, India

1. Introduction

In this new era electric power has become a fundamental part of the infrastructure of modern society, with most of daily activity is based on the assumption that the desired electric power is readily available for utilization. In the near future, electric supply to houses, offices, schools and factories is taken for granted. The complex power distribution system provides the required electricity to the customers.

The transfer of power from the generating stations to the consumers is known as an electric supply system (figure 1). It consists of three principal components, namely the generating station, transmission lines and distribution networks. The power is generated at favorable places which are quite far away from the customers. The power is produced and transmitted using a 3 phase 3 wire alternating current (A.C.) system and it is distributed using a 3 phase 4 wire A.C. system.

The distribution network components are the distribution substation, the primary feeder, distribution transformers, secondary distribution transformers, sectionalizing switches, tie switches and the services.

The network carries electricity from the transmission systems and delivers it to consumers at the load centres through a number of power lines (branches). Switching on and off of these power lines makes the power to flow in the power distribution network.

2. Power restoration in distribution system

The power distribution network can undergo outages, which may be forced or scheduled. Forced outages take place due to any faults in the network, whereas scheduled outages happen because of maintenance work. The various outages that occur in the distribution network are :

- Outage of the primary feeders
- Outage of the distribution transformers
- Outage of the distribution line

During outages, the supply of power is either partially or completely isolated from the feeder to the load centres. This deficit of power supply has to be the minimized. To achieve

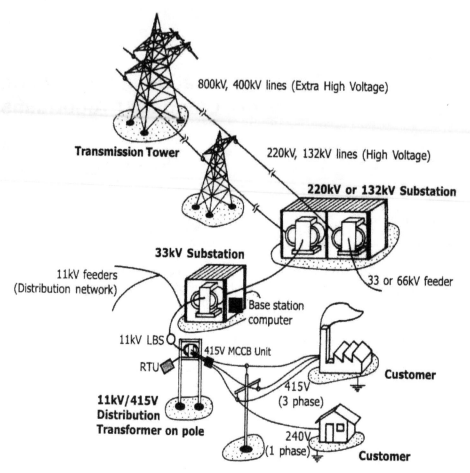

Fig. 1. Basic Electric Supply System

this goal, a proper switching sequence of power lines in the power distribution network is required. Sometimes, the load cannot be served to the customers; in which case, the loads are shed for the least priority customers.

The power distribution network in general, is built as an interconnected mesh network as shown in figure 2. Bus1, bus2 and bus3 indicate the feeder buses and the number in the circle indicates the load buses. The marking Si indicates the distribution branch i, that is used to transfer the power from one bus to another. Feeders in a distribution system have a mixture of types of loads, such as Very Important Person (VIP) & essential, industrial, commercial and domestic consumer loads. The peak load on feeders occurs at different times of the day, depending upon the type of load, making certain feeders to get heavily loaded and certain others to get lightly loaded. In such a practical situation, the reconfiguration based redistribution of the load amongst the feeders should attempt to evenly distribute the loads in the feeder.

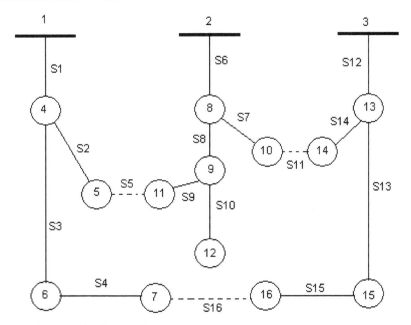

Fig. 2. Basic Power Distribution System

The branches in the power distribution network are normally configured radially for the effective coordination of their protective systems and such radiality is considered here. These networks are divided into a number of subsystems, which contain a number of normally closed switches (sectional switches) and normally open switches (tie line switches). These switches are operated during conditions of maintenance, dispatch and abnormalities. The aim of this switching operation is to reschedule the loads more effectively and improve the reliability of the distribution network. By changing the status of the switches, the topology of the power distribution network is reconfigured, and the resulting line currents and losses are redistributed, with a change in the bus voltage. These parameters of the network are obtained using Backward Sweep Power Flow algorithm.

In the case of an outage in any part of the system, it is imperative to restore the power system to an optimal target of the network configuration. The problem of obtaining a target network by switching is referred to as power system restoration. Power restoration after an outage usually refers to an emergency situation and the resultant plan should meet the following requirements:

- Restore as many loads as possible while considering priority customers
- Not cause violations in either engineering or operating constraints
- Outline a feasible sequence of operations to reach the final configuration
- Power balancing should be done
- Reached in a short time
- Radial network structure. This requirement is based on the feeder design for ease in fault location, isolation and protective device coordination.
- No components must be overloaded

The implementation of power restoration in a vast distribution system is thus a complicated combinatorial optimization problem because there are a great number of switches in the distribution system. It may take a long time using combinatorial optimization algorithm to reach a feasible restoration plan satisfying all practical constraints. An efficient way of achieving this would be to operate those switches that cause minimum loss and satisfy the voltage, current and other constraints. The major constraints to be satisfied in the distribution network are :

- Load allocation of the Feeder
- Voltage fluctuation
- Customer priority
- Reliability of the network
- Security of the network
- Distributed generation
- Harmonics due to intermittent switching
- Capacitor switching,
- Sudden increase of load and
- Failure of automated communication technology

Therefore, the dispatchers at many utilities tend to use their experience to narrow down and reach a proper restoration plan in a short period. This area has received a lot of attention by the researchers in the past three decades as evidenced by the number of publications (Sudhakar et. al 2011)

With the fast–paced changing technologies in the power industry, novel references addressing new technologies are being published. The automation of restoration of distribution power gained significance in the late eighties (Adibi and Kafka 1991). The state of the art methods used to solve the power system restoration for distribution system problems include Heuristic search (Morelato and Monticelli 1989), Expert system (Hotta et al 1990), and Knowledge based system (Matsumoto et al 1992). Due to the advancement in mathematics, new algorithms were developed to solve the restoration problem in distribution network. It mainly consisted of Artificial Neural Networks (Hoyong Kim et al 1993), Fuzzy Logic control (Han-Ching kuo and Yuan Yih Hsu 1993), Genetic Algorithm (Gregory Levitin et al 1995), Artificial Intelligence (Rahman 1993), Petri net (Fountas et al 1997), Tabu search (Toune 1998), Optimization (Nagata and Sasaki 2002), Ant colony search algorithm (Mohanty et al 2003) and Particle Swarm Optimization (Si-Qing Sheng et al 2009).

The main drawback faced in using the above methods, was the difficulty in identifying all the distribution branches used for the power to flow, after an outage for which predefined rules were used. To overcome this drawback, hybrid models such as fuzzy GA model (Ying-Tung Hsiao and Ching-Yang Chien 2000), was tried. To solve a complex combinatorial problem, the time required for solving the restoration problem using any of the above said methods is high. Now if hybrid models are used then the time required to obtain a solution is still higher. As a result it has become mandatory to identify the radial path of power flow with least mathematical efforts.

Most of the work reported focused on constraints like voltage limits, radiality and feeder capacity as the time required to obtain a restoration plan is more. To maintain these constraints, load shedding is done immediately. Load shedding option would imply loss of supply to essential loads such as medical facilities. If the time consumed is less; then the line losses and feeder capacity based on internal load division priority can be considered. To obtain the solution of the restoration problem without any iterative procedure, a graph theory based minimum spanning tree (MST) methodology proposed.

3. Graph theory

Graphs, the basic subject studied by the graph theory are abstract representations of a set of objects, where some pairs of objects are connected by links. The interconnected objects are represented mathematically as vertices, and the links that connect some pairs of vertices are called edges. Typically, a graph is depicted in a diagrammatic form as a set of dots for the vertices, joined by lines or curves called the edges. The vertices are also called nodes or points, and the edges are called lines.

A graph can be classified into two types namely an undirected and directed graph. A graph may be undirected, meaning that there is no distinction between the two vertices associated with each line, or its lines; or directed, meaning there is a distinction between one node and another. Table 1 shows the terminology for proceeding through the graph theory.

Term	Meaning
(a)	Vertex or node
(a)———(b)	The line joining two nodes or vertices is called an line. Since the line doesn't show the direction it is an undirected graph.
(a)——5——(b)	An line having a weight 5 being connected between the node 1 and node 2

Table 1. Symbols used in minimum spanning tree

Fig. 3. shows an example network with 6 nodes and 10 lines, which have their respective weights.

3.1 Minimum spanning tree

In the mathematical field of the graph theory, a spanning tree T of a connected, undirected graph G is a tree composed of all the vertices and some (or perhaps all) of the lines of G. That is, every node lies in the tree, but no cycles (or loops) are formed. A spanning tree of a connected graph G can also be defined as a maximal set of lines of G that contains no cycle, or as a minimal set of lines that connect all the vertices. For a connected graph with V nodes, any spanning tree will have the V-1 lines.

Given a connected, undirected graph, a spanning tree of that graph is a subgraph which is a tree and connects all the vertices together. A weight is assigned to each line, whose value represents how unfavorable it is for the considered task. Individual weights of lines in a spanning tree decide its weight. The total sum of all the weights of the lines in a particular spanning tree is its weight.

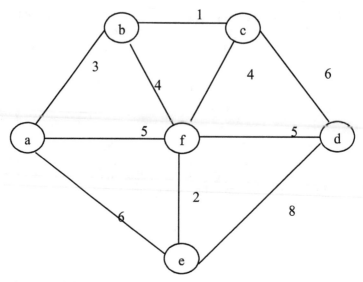

Fig. 3. Example network for MST problem

A minimum spanning tree (MST) or minimum weight spanning tree is then a spanning tree with a weight less than or equal to the weight of every other spanning tree. This concept is often used in routing. The algorithms to find MST, a graph search algorithm that solves the single-source shortest path problem for a graph with non–negative line path weights, produces a shortest path tree. For a given source node in the graph, the algorithm finds the path with the lowest weight (i.e., the shortest path) between that node and every other node. For example, if the nodes of the graph represent cities, and the line path weights represent the driving distances between pairs of cities connected by a direct road then MST can be used to find the shortest route between one city and all other cities. The main advantage of using the MST is that the optimum solution is obtained in a single stage.

3.2 Algorithms for finding MST

This section presents the proposed MST algorithm. Each algorithm is explained with a common example. A single graph can have many different spanning trees. If an exhaustive search approach to construct an MST is tried, two serious drawbacks arise. First, the number of spanning trees grows exponentially with the graph size; second, generating all the spanning trees for a given graph is not easy. To overcome these drawbacks the MST algorithms are proposed. The MST algorithms are further classified as line based MST algorithms and node based MST algorithms. They are

1. Line based MST algorithms
    i.   Kruskal's algorithm
    ii.   Reverse Delete algorithm
2. Node based MST algorithms
    i.   Prim's algorithm
    ii.   Dijkstra's algorithm

3.3 Kruskal's algorithm

Kruskal's algorithm is an algorithm in graph theory presented in 1956 (Anany Levitin (2009)) that finds a minimum spanning tree for a connected weighted graph. This means that it finds a subgraph of the lines that form a tree. It includes every node of the network and the total weight of all the lines in the tree is minimized. To implement the Kruskal's algorithm, two conditions have to be satisfied. First the weight of the lines in a graph is arranged in the increasing order. Second an empty subgraph T (called Traversal matrix) is created.

Then, the algorithm considers a line, based on the order of increasing weight. If a line (u, v) $(u, v$ are the starting & ending node of a line) does not form a cycle along with the existing lines in the subgraph (T) then the line (u, v) is added to the subgraph (T). Then the line (u, v) is discarded. If a line is added to the subgraph then the counter is incremented. It is checked that the counter value is equal to V–1, where V is the number of nodes. If it is true, the procedure is stopped; otherwise, the process continues.

The Kruskal's algorithm is applied to the example network as shown in figure 3. The step by step procedure of the algorithm is discussed by Sudhakar et al (2011). The resultant traversal matrix with a weight of 15 is

$$T = [b\ c$$
$$f\ e$$
$$b\ a$$
$$b\ f$$
$$f\ d]$$

At the termination of the algorithm, the traversal matrix forms a MST of the graph. Based on this process, a pseudo code of the algorithm is developed and is as follows,

Pseudocode for the Kruskal's algorithm:

Sort E in increasing order of edge weights and the sorted edges are in A.

- Initialize the set of tree edges and its size, T.
- Counter $\leftarrow 0$
- While counter < $|V|$-1
- If $T\ U\ (u, v)$ is acyclic
- $T \leftarrow T\ U\ (u, v)$
- Counter $\leftarrow$ Counter+1

Return T

3.4 Reverse Delete algorithm

The Reverse-Delete algorithm is an algorithm in graph theory used to obtain a minimum spanning tree from a given connected, line-weighed graph. The Reverse-Delete algorithm starts with the original graph and deletes lines from it. The Reverse-Delete algorithm ensures connectivity in the graph before deletion. Since the algorithm only deletes lines when it does not disconnect the graph, any line removed by the algorithm forms a cycle prior to the deletion. Since the algorithm starts from the maximum weighted line and continues in descending order, the line removed from any cycle is the maximum weighted line in that cycle. Therefore, according to the definition of a minimum spanning tree, the lines removed by the algorithm are not in any minimum spanning tree.

The Reverse-Delete algorithm is applied to the example network as shown in figure 3. The step by step procedure of the algorithm is discussed by Sudhakar et al (2011). Thus, the minimum weight for traversing the graph is 15 and the resultant network is shown in figure 4.

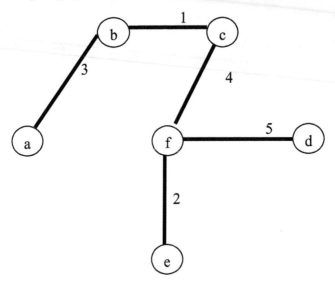

Fig. 4. Final Result of Reverse-Delete algorithm

Based on this procedure a pseudo code of the algorithm is formed as follows :

Pseudocode for the Reverse-Delete algorithm:
Function Reverse Delete (edges [T])
Sort T in decreasing order
Define an index i ← 0
    while i < size(T)
    Define edge temp ←T[i]
    delete T[i]
 if temp.v1 and temp.v2 are not connected to the tree
        T[i] ← temp
      i ← i+1
return edges[T]

3.5 Prim's algorithm

The algorithm was developed in 1930 by the Czech mathematician, Vojtech Jarnik, and later independently, by the computer scientist, Robert C. Prim, in 1957 (Anany Levitin (2009)).

This algorithm finds a subset of the nodes that form a tree that includes every node, where the total weight of all the lines in the tree is minimized.

The Prim's algorithm constructs an MST through a sequence of expanding subtrees. The initial subtree in such a sequence consists of a single node selected arbitrarily from the set $V-V_T$ of the graph's nodes. In the following steps, the current tree expands, by simply getting attached to a nearer node with less weight. The algorithm stops, after all the graph's nodes have been included in the tree being constructed. Since the algorithm expands a tree by exactly one node on each of its steps, the total number of such steps is $V-1$, where V is the number of nodes in the graph.

The nature of the Prim's algorithm makes it necessary to provide two data values for every other unselected node. The data values are provided through two arguments : first will be the unselected node's $(V-V_T)$ connectivity to the currently selected node (V_T). It will be nil '–' if no connectivity exists. The second entry (distance label) will be the respective weight. If there is no connection then the value will be ∞. With such labels, the smallest distance label in the set $V - V_T$, is selected and added in the selected nodes list.

After a node $e*$ is identified which is to be added to the tree, the following operations have to be performed:

- Move $e*$ from the set $V-V_T$ to the set of selected nodes V_T.
- Based on the nodes in set V_T, the weights of the node in $V-V_T$ are updated.
- For each remaining node u in $V-V_T$, $e*$ is selected which has the minimum weight
- The $e*$ is the next node to be added to the current tree T and the node $e*$ is added in V_T

The Prim's algorithm is applied to the example network as shown in figure 3. The step by step procedure of the algorithm is discussed by Sudhakar et al (2011). Thus, the minimum weight for traversing the graph is 15 and the resultant network is shown in figure 5.

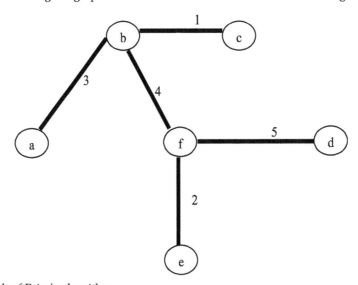

Fig. 5. Result of Prim's algorithm

Based on this procedure a pseudo code of the algorithm is formed as follows :

ALGORITHM *Prim(G)*
> //Prim's algorithm for constructing an MST
> //Input: A weighted connected graph G = *(V, E)*//V is node; E is line
> //Output: E_T, the set of lines composing the MST of G
> //the set of tree nodes can be initialized with any node
> $E_T \leftarrow \emptyset$
> *for i=1 to (V-1) do*
> find a minimum-weight line $e^* = (v^*, u^*)$ among all the lines *(v,u)*
> such that *v* is in V_T and *u* is in $V - V_T$
> $$V_T \leftarrow V_T \cup \{u\}$$
> $$E_T \leftarrow E_T \cup \{v\}$$
> return E_T

3.6 Dijkstra's algorithm

The Dijkstra's algorithm, was conceived by the Dutch computer scientist Edsger Dijkstra in 1959 (Anany Levitin (2009)). This algorithm uses traversal matrix [T]. It indicates the distance from a single node to all the other nodes including it. In the considered example of figure 3, when node 'a' is considered first, its distance from it and all nodes from a through f will form the initial entries of [T]. The same procedure is repeated till all the nodes are selected and the final network with a total weight of 20 as shown in figure 6. algorithm is applied to the example network as shown in figure 3. The step by step procedure of the algorithm is discussed by Sudhakar et al (2010). Thus, the minimum weight for traversing the graph is 20 and the resultant network is shown in figure 6.

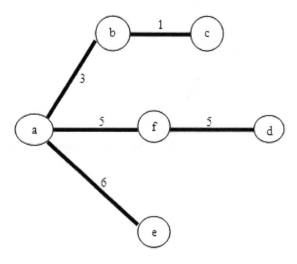

Fig. 6. Resultant network of Dijkstra's algorithm

Based on this modus operandi, a pseudo code of the algorithm is developed as follows,

Pseudocode for the Dijkstra's algorithm:
function Dijkstra(*Graph, source*):
// Initializations
for each node *v* in *Graph*:
// Unknown distance function from s to v
 dist[*v*] := infinity previous[*v*] := undefined
// Distance from s to s
 dist[*source*] := 0
// Set of all unvisited vertices
 Q := copy(*Graph*)
// The main loop
while *Q* **is not** empty:
// Remove best node from the priority queue;
// returns to the source after first step
 u := extract_min(*Q*)
 for each neighbor *v* of *u*:
  alt = dist[*u*] + length(*u, v*)
// Relax (u,v)
   if *alt* < dist[*v*]
   dist[*v*] := *alt*
   previous[*v*] := *u*

3.7 Comparison of the results

The four algorithms are applied to the example network of figure 3, and the results are tabulated in table 2.

Algorithm	Total weight
Prim's algorithm	15
kruskal's algorithm	15
Dijkstra's algorithm	20
Reverse Delete algorithm	15

Table 2. Comparison of MST algorithms for example network

In the case of the Dijkstra's algorithm, the total weight of the resultant network is more than that of the other three algorithms; but for the case of node 'f' from the starting node 'a' the total weight using the Dijkstra's algorithm is 5 which is smaller than the value 7 obtained using the other three algorithms. As a result we cannot neglect the Dijkstra's algorithm as an inefficient method.

3.8 MST applications

Graph theory based on MST approach has been discussed in various papers for various applications. Lin Ming Jin & Shu Park Chan (1989) presents an algorithm for finding the shortest path for power routing (DC) between two nodes in an electrical network used in the airlines. Hiroyuki Mork, Senji Tsuzuki (1991) presents mathematically MST for network topological observability analysis. Shun Lin Su et al (1994) dealt with application of the MST for finding the connectivity in VLSI circuits. Cavellucci & Lyra (1997), presented the minimization of energy losses in distribution systems by applying a general search method to a Brazilian power network. Here outages were not considered as an important factor. Michel Barbehenn (1998) discusses about the application of Dijkstra's algorithm for various applications like airline electrical networks. Ali Shatnawi et al (1999) indicates the use of Floyd – Warshall's based MST to find the time scheduling in the data flow graph of a DSP. Partricia Amancio Vargas (2002) uses the learning classifier system for loss minimization in a power system. Kaigui Xie (2003) calculates the reliability index of radial network using forward search method of MST. TianTian CAi & Qian Ai (2005) discusses the depth first search method used to find the MST for the optimal placement of the PMU devices in the power system. Distribution reconfiguration algorithm, named Core Schema Genetic Shortest-path Algorithm (CSGSA) proposed by Yixin Yu & Jianzhong Wu (2002) is based on the weights calculation method for each load condition based on line losses.

The above survey highlighted the extension of the application of graph theory for MV power distribution AC system, which has been attempted at this juncture. Here, the mathematical formulation of Yixin Yu has been applied to a PDN wherein distribution branch outages have been fully addressed. Thus to obtain the restoration plan quickly without any iterative procedure, a graph theory based methodology using MST algorithms is proposed here. Four algorithms based on graph theory are used to restructure the PDN by considering the distribution branch outage which forms the major contribution of this work.

The MST algorithm identifies all the possible paths for the power to flow and obtains only one solution. In a single iteration the MST algorithm overcome the radiality constraint. Since, MST algorithm gives a path of minimum impedance, the line losses will be minimum with the result no separate loss minimization procedure is required here. The solution of MST algorithm minimizes the solution time and as a result loss minimization and load shedding with internal priorities are included in the proposed work. Thus, in a minimum time a power system restoration solution is obtained, which will not lead to cascaded outage.

4. Restoration problem

An outage degrades the most important function of an electrical system, that of supplying the customers, and thus has a radical influence on the operating objectives. Whenever power supply interruption occurs in distribution systems due to an outage, it is imperative to bring back the system promptly to its initial state or to an optimal target network, by switching operations. The problem of obtaining a target network is called as power system restoration, has two prime objectives (a) the number of customers with a restored supply should be the largest possible and (b) the restoration should be accomplished as quickly as possible.

In this section, the network reconfiguration problem for service restoration is discussed in detail. The system is represented on a per phase basis and the load along a feeder section is represented as constant P, Q loads placed at the end of the lines. It is assumed that every switch is associated with a line in the system. The network reconfiguration problem for loss reduction involves the load transfer between the feeders or substations by changing the position of the switches. The radial configuration corresponds to a 'spanning tree' of a graph representing the network topology.

Given a graph, find a spanning tree such that the problem formulation of the restoration problem is given here

Objective Function is to

Find the Optimal Power Path

With the following constraints:

i.   Maximize the power restored to the isolated area,

$$\max f = \sum_{K \in B} L_K X_K \qquad (1)$$

Where L_K is the load at bus K

B is the total number of buses

X_K is the decision variable

The position of the sectional and tie line switches is considered here as the binary decision variable. This variable decides whether the switch is open or closed. The binary decision variable, X_K is defined as

$$X_K = 1 \ \text{(if the K}^\text{th}\text{ switch is closed)}$$

$$= 0 \ \text{(if the K}^\text{th}\text{ switch is open)} \qquad (2)$$

If there are m switches in the initial network, there will be 2^m on / off switching options. That is, for a two switch network, 2^2 switching options are possible [(0,0), (0,1), (1,0), (1,1)]. The resultant decision vector is given by

$$X = [X_1, X_2, X_3 \ldots\ldots\ldots\ldots\ldots X_m]. \qquad (3)$$

For the two switch network the decision vector can be represented as $X = (0,0)$, $X = (0,1)$, $X = (1,0)$ and $X = (1,1)$.

ii. Voltage limits : For each bus bar, the voltage constraints have upper and lower limits. A general expression for voltage constraints would be

$$\left| V_{\min j} \right| \leq \left| V_j \right| \leq \left| V_{\max j} \right| \qquad (4)$$

where $\left| V_j \right|$ is the voltage at j node, $\left| V_{\min j} \right|$ is the minimum permitted voltage at node j, $\left| V_{\max j} \right|$ is the maximum permitted voltage at node j, and j belongs to a set of buses where voltage constraints are observed.

iii. Radiality : It is a condition in power distribution network that only one feeder bus feeds a load bus.

iv. Loading constraints : A general expression for loading constraints would be

$$L_i < L_{MP \, i} \qquad (5)$$

where L_i is the loading of the network element i, $L_{MP\ i}$ is the maximum permitted loading of the network element i, and i belongs to a set of protected network elements. These protected network elements, are normally lines and feeders that can be protected by actual protection devices, or can be algorithmically protected. However, the fact is that, the considered feeder is not capable of supplying the whole load and hence, cannot be used further for problem solving.

v.   Line losses: The total power losses of the network should be minimum.
vi. Feeder capacity: The total capacity of the feeder should not be violated, and
vii. Priority of customers: As the priority of each service zone is determined in advance, which customers would be restored can be determined according to the $L_{MP\ i}$.

This is a combinatorial optimization problem, since the solution involves the consideration of all possible spanning trees.

5. Application of the proposed MST algorithms for restoration problem

MST algorithm is a graph search algorithm that solves the single-source shortest path problem for a graph with non–negative weights, producing a shortest path tree. For a given source vertex (node) in the graph, the algorithm finds the path with lowest impedance (i.e. the shortest path) between the source node and every other node.

In order to achieve a maximum amount of power restored in a radial distribution system, the aim is to identify the appropriate switching options, which consists of all the buses in the network. In the proposed method, the distribution system is considered with all its laterals simultaneously, instead of determining the switching options on loop by loop basis. In applying the graph theory the buses and the feeders are considered as the node, the distribution line is considered as line and the impedance of the distribution line is considered as weight of the line. With this consideration the proposed graph theory based algorithm for the distribution system is :

Step 1.   Initial power flow path is stored
Step 2.   Get the input data about the amount of loads at each bus, the feeder capacity and the current status of all the lines
Step 3.   In case of any outage remove those data from the input data file
Step 4.   Check for multi feeder or single feeder
Step 5.   If it is single feeder go for step 7
Step 6.   If it is multi feeder, then enter the number of feeders
Step 7.   Get the MST for the feeders
Step 8.   Perform the load flow for the resultant network
Step 9.   Check for multi feeder or single feeder
Step 10. If it is single feeder go to step 13
Step 11. Check for feeder overloading condition. If the feeders are overloaded, then load transferring can be done. If load transferring is not possible then perform load shedding. Otherwise if the feeders are not overloaded proceed to step 13.
Step 12. Perform the load flow for the modified conditions
Step 13. Check for over voltage condition. If the voltage limits are violated then perform load shedding otherwise proceed to step 15.
Step 14. Perform the load flow for the modified conditions

Step 15. Check for over current condition. If the current capacity are violated then perform load shedding otherwise proceed to step 17.

Step 16. Perform the load flow for the modified conditions

Step 17. If all the constraints are satisfied then display the optimal switching sequence

By applying this methodology objective function is solved and the constraints are satisfied. The result of step 7 finds a path for the power to flow after an outage, which satisfies the objective function. This step also satisfies the constraints viz., maximize the power restored to the isolated area, radiality and line losses. The power restored to the isolated area is maximum as the MST obtained after step 7 has all the possible buses available in the network. So all the loads connected to these buses will receive the power. The MST does not allows any closed loop in the network, as a result the MST obtained after step 7 will not have any closed loops, so the radiality constraint is satisfied. The MST network will have a minimum impedance value because the MST is obtained by considering the impedance value of each distribution line. Then the losses of the network will be minimum.

The feeder capacity constraint is mainly applicable for the multi feeder networks. In the case of the single feeder network, the loads are rearranged in the same network whereas in the multi feeder network there are some conditions that the loads of a feeder are transferred to the next feeder nearby. This condition is checked in the step 10. If there is any feeder overloading then load transferring is done. If load transferring is also not possible then load shedding is done based on the priority of the customer's constraint.

Step 13 checks the voltage limit condition and step 15 checks the loading constraint of the network. The maximum allowable limits are based on the network considered. Thus the proposed methodology satisfies all the above said objective function and the constraint.

The proposed methodology is not an iterative procedure, it calculates the amount of the load to be shed at each bus or load transferring between feeder (in the case of multi feeders) three times to satisfy the constraints. The load shed or load transferring has to be performed three times because the MST obtained by step 7 is based on the impedance minimization; it means the load has no influence on the solution. To bring in the effect of the load conditions only the load shedding or load transferring is done. Each time the amount of load to be shed is calculated at each bus and finally all the loads are shed simultaneously.

6. Results of the proposed MST algorithms for restoration problem

The original configuration of 33–bus test distribution network shown in figure 7 has a total load capacity of 3.525 MW and 2.3 Mvar. The network consists of 33 buses and 37 branches, where branches S1–S32 and S33–S37 indicate the sectional and tie lines switch respectively. The total impedance of the network for the original configuration having sectional lines is 21.8744+ j 18.1456 and the loss of the network 0.1869MW and 0.1240Mvar.

For the given network, the proposed methodology using the four MST algorithms is applied and the switch that should be kept open under normal conditions is tabulated in Table 3. The table indicates the switches those act as tie switches, their total impedance of the resultant network, loss of the network after applying the Kruskal's algorithm, Reverse delete algorithm, Prim's algorithm & Dijkstra's algorithm and minimum p.u. voltage of the network.

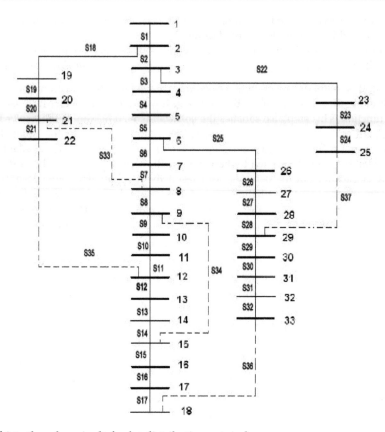

Fig. 7. Thirty three bus single feeder distribution network

Algorithm	Tie Switches					Total Impedance		Real power loss	Reactive power loss	Minimum bus Voltage
						R	X			
						Ω	Ω	MW	Mvar	p.u.
KRUSKAL'S	S16	S27	S33	S34	S35	20.52	16.49	0.1780	0.1230	0.922
REVERSE DELETE	S16	S27	S33	S34	S35	20.52	16.49	0.1780	0.1230	0.922
PRIM'S	S16	S27	S33	S34	S35	20.52	16.49	0.1780	0.1230	0.922
DIJKSTRA'S	S9	S14	S16	S28	S33	26.15	22.46	0.1671	0.1192	0.928

Table 3. Switches that are open and the impedance of the network

From the table it is observed that the resultant losses of the network obtained using the MST algorithms for the normal condition are less when compared to that of the original configuration. In the case of Dijkstra's algorithm it is noted that the impedance value is higher than the total impedance obtained using other three MST algorithm and the initial configuration. The value of the loss is less when compared to other minimum spanning tree algorithms and the initial network.

For the same network hybrid GA (Jizhong Zhu and Chang (1998)) and heuristic search method (Shirmohammadi and Hong (1989)) are applied and their results are tabulated in Table 4.

Algorithm	Tie Switches					Total Impedance (Ω)		Real power loss	Reactive power loss	Minimum bus Voltage
						R	X	MW	Mvar	p.u.
REFINED GA	S7	S10	S14	S36	S37	25.38	22.31	0.2007	0.1776	0.883
HEURISTIC METHOD	S7	S9	S14	S32	S37	24.39	21.61	0.1984	0.1760	0.887

Table 4. Switches that are open and the impedance of the network

Using the proposed methodology for a single line outage in 33 – bus network, the results are obtained and tabulated in Table 5.

OUTAGE IN LINE	SWITCHES THAT ARE OPEN				
S1	Power system cannot be restored				
S2	S2	S8	S24	S32	S34
S3	S3	S7	S9	S14	S16
S4	S4	S7	S9	S14	S16
S5	S5	S7	S9	S14	S16
S6	S6	S9	S13	S16	S28
S7	S7	S9	S16	S28	S34
S8	S8	S15	S28	S33	S34
S9	S9	S14	S16	S28	S33
S10	S10	S14	S16	S28	S33
S11	S11	S14	S16	S28	S33
S12	S12	S9	S16	S28	S33
S13	S13	S9	S16	S28	S33
S14	S14	S9	S16	S28	S33
S15	S15	S9	S14	S28	S33
S16	S16	S9	S14	S28	S33
S17	S17	S9	S14	S28	S33
S18	S18	S13	S16	S28	S35
S19	S19	S13	S16	S28	S35
S20	S20	S13	S16	S28	S35
S21	S21	S14	S16	S28	S33
S22	S22	S9	S14	S16	S33
S23	S23	S9	S14	S16	S33
S24	S24	S9	S14	S16	S33
S25	S25	S9	S14	S16	S33
S26	S26	S9	S14	S16	S33
S27	S27	S9	S14	S16	S33
S28	S28	S9	S14	S16	S33
S29	S29	S9	S14	S28	S33
S30	S30	S9	S14	S28	S33
S31	S31	S9	S14	S28	S33
S32	S32	S9	S14	S28	S33

Table 5. Result for single line outage in 33 bus network

6.1 Hardware implementation of the proposed MST algorithms for restoration problem

For the real time application in the automated world the developed program has to be used with hardware. So the developed methodology is implemented using Verilog HDL. Verilog is a hardware description language (HDL). A HDL is a language used to describe a digital system, for example, a network switch, or any memory or a single flip flop. This means that by using a HDL, one can describe any hardware at any level.

The proposed methodology indicates the ON and OFF status of the switch. In this hardware the ON and OFF status of a line is denoted by LOGIC 1 and LOGIC 0 respectively. The status of all the lines is denoted through on-board LEDs which is interfaced with the FPGA kit. The I/O pins of the FPGA chip are configured and port mapped accordingly. The Verilog program for 33 – bus single feeder system is implemented in Verilog through SPARTAN 3 FPGA (Field Programmable Gate Array) kit (Figure 8). It uses a XILINX XC3S400 chip for processing. The XC3S400 FPGA chip is embedded in a kit with peripheral ICs and components for research and development purpose.

Fig. 8. XILINX SPARTAN3 FPGA Kit

A Verilog program is written using XILINX ISE software to program the FPGA and display the status of lines as output based on the outage line which is given as input. The chip is programmed using XILINX ISE navigator tool. The FPGA kit used in this project consists of sixteen input switches and sixteen output LEDs (Figure 9). In addition the kit also has a 34 pin FRC and 60 pin FRC connector for 94 I/Os. Using these additional pins, the status of other switches (S17 – S37) are indicated in a bread board (Figure 10).

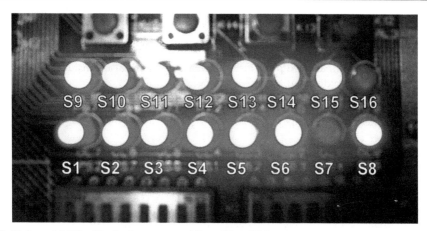

Fig. 9. Onboard LEDs displaying status of lines S1 to S16

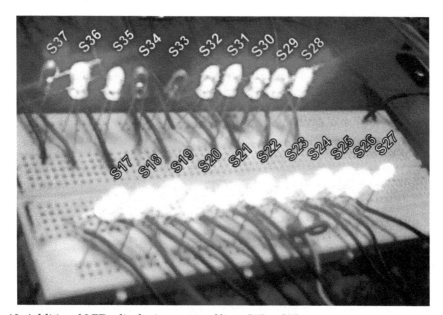

Fig. 10. Additional LEDs displaying status of lines S17 to S37

7. Conclusion

A feeder reconfiguration method using MST for service restoration of radial distribution system is presented. From the important observations of the present study it could be concluded that:

- The out-of-service area is reduced to the maximum by the developed MST methodology
- The power losses of distribution systems are reduced by proper feeder reconfiguration.

- In addition to power–loss reduction, the voltage profile is improved by the proposed method.
- Based on the methodology the feeder loads and load flow are performed each time, so that the effect of unbalanced power distribution network is also considered.
- It can be applied to distribution network of any size.

Test results obtained indicate that, this method results in better restoration plan when compared to other reference papers.

8. References

Adibi, M. M. and Kafka, R. J. (1991). *Power System Restoration Issues*, IEEE Computer Applications in Power, Vol. 4, No. 2, pp. 19-24.

Ali Shatnawi, M O Ahmad and M N Swamy, (1999) *Scheduling of DSP data flow graphs onto multiprocessor for maximum throughput*, IEEE 1999, pp 386 – 389.

Cavellucci and Lyra, (1997), *Minimization of energy losses in electric power distribution system by intelligent search strategies*, International transaction in operational research, vol. 4, no. 1, pp 23 – 33.

Fountas, N. A. Hatziargyriou, N. D. and Valavanisl, K. P. (1997). *Hierarchical Time Extended Petri Nets As A Generic Tool For Power System restoration*, IEEE Transactions on Power Systems, Vol. 12, No. 2, pp. 837-843.

Gregory Levitin, Shmuel Mazal–Tov and David Elmakis, (1995). *Genetic algorithm for optimal sectionalizing in radial distribution systems with alternative supply*, Electric Power Systems Research, Vol. 35, pp. 149-155.

Han – Ching kuo and Yuan Yih Hsu, (1993). *Distribution System Load Estimation and Service Restoration using a Fuzzy Set Approach*, IEEE Transactions on Power Delivery, Vol. 8, No. 4, pp. 1950-1957.

Hiroyuki Mork and Senji Tsuzuki, (1991), *A fast method for topological observability analysis using minimum spanning tree technique*, IEEE Transaction on power system vol. 6, no. 2, May 1991, pp 491 – 500.

Hotta, K., Nomura, H., Takemoto, H., Suzuki, K., Nakamura, S. and Fukui, S. (1990). *The Implementation of a Real-Time Expert System for a Restoration Guide in a Dispatching Center*, IEEE Transactions on Power Systems, Vol. 5, No. 3, pp. 1032-1038.

Hoyong Kim, Yunseok ko and Kyung–Hee Jung, (1993), *Artificial Neural-Network based Feeder Reconfiguration for Loss Reduction in Distribution Systems*, IEEE Transactions on Power Systems, Vol. 8, No. 3, pp. 1356-1366.

Kaigui Xie, Jiaqi Zhou and R Billinton, (2003), *Reliability evaluation algorithm for complex medium voltage electrical distribution networks based on the shortest path*, IEE Proc.-Gener. Transm. Distrib., vol. 150, no. 6, November 2003, pp 686 – 690

Lin Ming Jin and Shu Park Chan, (1989), *An electrical method for finding suboptimal routes*, ISCAS'89 IEEE pp 935 – 938.

Matsumoto, K., Sakaguchi, T., Kafka, R. J. and Adibi, M. M., (1992), *Knowledge-Based Systems as Operational Aids in Power System Restoration*, Proceedings of the IEEE, Vol. 80, No. 5, pp. 689-697.

Michel Barbehenn, (1998), *A note on the complexity of Dijkstra's algorithm for graphs with weighted vertices*, IEEE Transactions on computers, vol. 41, no 2, feb 1998, pp 263.

Mohanty, I., Kalita, J., Das, S., Pahwa, A. and Buehler, E., (2003), *Ant algorithms for the optimal restoration of distribution feeders during cold load pickup*, Proceedings of the 2003 Swarm Intelligence Symposium, SIS '03, IEEE, pp. 132-137.

Morelato, A. L. and Monticelli, A. (1989) *Heuristic Search Approach to Distribution System Restoration*, IEEE Transactions on Power Delivery, Vol. 4, No. 4, pp. 2235-2241.

Nagata, T. and Sasaki, H., (2002), *A Multi-Agent Approach to Power System Restoration*, IEEE transactions on power systems, Vol. 17, No. 2, pp. 457- 462.

Partricia Amancio Vargas, Christiano Lyra Filho and Fernanado J Von Zuben, (2002), *On line approach for loss reduction in electric power distribution networks using learner classifier systems*, Springer, pp 181 – 196.

Rahman, S., (1993), *Artificial intelligence in electric power systems – a survey of the Japanese industry*, IEEE Transactions on Power System, Vol. 8, No. 3, pp. 1211-1218.

Shun Lin Su, Charles H Barry and Chi Yuan Lo, (1994) *A space efficient short finding algorithms*, IEEE Transactions on computer aided design of integrated circuit & systems, vol. 13, no. 8, August 1994, pp 1065 – 1068.

Si – Qing Sheng, Yun Cao and Yu Yao, (2009), *Distribution Network Reconfiguration Based on Particle Swarm Optimization and Chaos Searching"*, Asia-Pacific Power and Energy Engineering Conference, APPEEC 2009, IEEE, pp. 1-4

Sudhakar T D, et. Al, (2010), *"Prim's Algorithm for Loss Minimization and Service Restoration in Distribution Networks"*, International Journal of Electrical and Computer Engineering (IJEC), Volume 2, Number 1 (2010), pp. 43 – 62

Sudhakar T D, et. Al, (2010), *"A Graph Theory - Based Distribution Feeder Reconfiguration for Service Restoration"*, International Journal of Power and Energy Systems, Vol. 30, No. 3, 2010, pp 161 – 168

Sudhakar T D, et. Al, (2011), *"Power System Reconfiguration based on Kruskal's Algorithm"* IEEE conference ICEES 2011 Volume 1 Page No 234 – 240

Sudhakar T D, et. Al, (2011), *"Power System Restoration using Reverse Delete Algorithm Implemented in FPGA"*, Second IET International Conference on Sustainable Energy and Intelligent System, Page : 373 – 378

Sudhakar T D, et. Al, (2011), *"Restoration of Power Distribution Network – A Bibliographical Survey"*, European Transactions on Electrical Power, Vol. 21, pp 635 – 655

TianTian CAi and Qian Ai, (2005), *Research of PMU optimal placement in power systems"*, International conference on system theory and scientific computation, pp 38 – 43, ISBN 960-845735-1

Toune, S., Fudo, H., Genji, T., Fukuyama, Y. and Nakanishi, Y., (1998), *A reactive tabu search for service restoration in electric power distribution systems"* The 1998 IEEE International Conference on Evolutionary Computation Proceedings, IEEE World Congress on Computational Intelligence, pp. 763-768.

Ying – Tung Hsiao and Ching – Yang Chien, (2000), *Enhancement of Restoration Service in Distribution Systems Using a Combination Fuzzy – GA Method*, IEEE Transactions On Power Systems, Vol. 15, No. 4, pp. 1394-1400.

Yixin Yu, and Jianzhong Wu, (2002), *Loads Combination Method Based Core Schema Genetic Shortest-path Algorithm For Distribution Network Reconfiguration*, 2002 IEEE, pp 1729 – 1733.

Applications of Graphical Clustering Algorithms in Genome Wide Association Mapping

K.J. Abraham[1]* and Rohan Fernando[2]
[1]*Programa de Pós Graduação em Genética, Faculdade de Medicina de Ribeirão Preto,
Universidade de São Paulo*
[2]*Dept. of Animal Science, Iowa State Univ*
[1]*Ribeirão Preto SP, Brazil*
[2]*Ames IA USA*

1. Introduction

The field of statistical genetics has been the area of a great deal of active research in recent years; due in part to dramatic advances in sequencing technology which has led to vast amounts of genomic data becoming available at lower and lower costs. The data from these sequencing efforts is not only copious, but is also characterized by significant levels of experimental noise. Processing data of this nature to draw statistically significant inferences requires dealing with a number of challenges, both statistical and algorithmic. In this chapter we will not discuss the very substantial statistical issues which arise in extracting genome sequences, but rather will focus on computational and algorithmic issues which arise in drawing biological inferences once the sequence is known. Event though our discussion will be oriented more towards applications of graph theory, it is worth keeping in mind that statistical considerations will still play a role due the intrinsically probabilistic nature of Mendelian Genetics as well as finite sample sizes. We will first begin by reviewing earlier well known work which will serve to illustrate the utility of graph theoretical concepts in dealing with genomic data. The data at our disposal are assumed to consist of observations of at a large number of locations (tens of thousands or possibly much more) on multiple chromosomes for a collections of individuals, or plants or animals; for now we assume that these individuals are related with known parent offpsring information. We restrict our attention to species which have just two chromosomes, but much of what we will discuss can be generalized to species with more than two chromosomes although the computational implementation could be challenging. At any locus (precise location on a specified chromosome) we assume that there are two or more possible alleles in the population, the precise number is assumed to differ from locus to locus. The number of possible observable genotypes at each locus will thus also vary from locus to locus. In a population of related individuals with known parent offspring relations between individuals (*i.e.* pedigree) it is possible to predict the probability for an offspring to receive a given allele from a parent based on Mendelian Genetics. If we represent a given locus for a given individual by a vertex, we can assign multiple possible states to each vertex depending on how many genotypes are possible. Since genetic information flows from parents to offspring

---

*Bolsista CAPES/Brasil

we can construct a directed graph where the arrows flow from parental vertices to offspring vertices. The indegree of each vertex is maximally two, while the outdegree will depend on how many offspring an individual has. We thus have a directed graph where each vertex has multiple states (genotypes or alleles) associated with it. Since individuals cannot be their own ancestors, the graph is acyclic as well as being directed. With each edge we can associate a transmission probability which is determined by Mendelian Genetics; in addition there is a Markov Field Property involved as conditional on parental genotypes offspring genotypes are independent of all other ancestral genotypes and sib genotypes. We thus have all the ingredients of a Bayesian Network. Many important problems relating to genetic inference from data on pedigrees have been formulated in the language of Bayesian Networks (Fishelson & Geiger, 2002); (Fishelson & Geiger, 2004); finding exact and approximate solutions to these problems particulary on large data sets has led to the development of very sophisticated algorithms which we will not discuss here. The key feature underlying the data is that it consists of a potentially large but discrete number of observations. These observations have a very complex correlational structure, some of the observations are heavily correlated (*eg.* the genotypes of parents and offspring at the same locus) while others may be very loosely correlated (*eg.* the genotypes of individuals and ancestors going back several generations). The discrete nature of the data points permits us to assign a vertex in a graph to each data point while the edge structure (which arises from the pedigree structure) is a reflection of the association between data points; seen in this light the use of a graph theoretical formulation is quite natural. In what follows, the data under consideration will have the same features, suggesting the use of graphical models but our discussion will focus more on the use of undirected graphical models.

In order to motivate the application of undirected graphical models, we consider two fixed loci on the same chromosome in a population of individuals. It is frequently observed that the joint distribution of alleles at loci which are in close physical proximity on the same chromosome, is not uniform. More specifically, if there are two alleles at one locus A or a and B or b at another locus, then the probability of finding allele B in some arbitrary individual in a population may depend on the which allele (A or a) is present at the other locus. In the language of probabilities $\mathcal{P}(A, B) \neq \mathcal{P}(A)\mathcal{P}(B)$ for certain pairs of loci; this is the phenomenon of linkage disequilibrium (LD) (Weir, 1996). The extent and statistical significance of the non-randomness of the alleleic association can be quantified by analyzing a (3×3) contingency table whose entries are genotype counts. If there are just two of the three possible genotypes present, or if the population from which the unrelated individuals are sampled is subject to certain other constraints, it becomes possible to estimate the linkage disequilibrium between each pair of distinct markers using just a (2×2) contingency table, alternatively the non-randomness in the asociation between the alleles depends one just one function of the allele counts. What is actually computed is just the sample LD, the standard errors on the LD will depend inversely on the size of the population. The magnitude of the observed LD can vary quite dramatically depending on which loci are being compared, large LD is very much more common among loci close together than loci on different chromosomes. Furthermore, while LD tends to decrease as the distance between loci increases, the decrease is often neither uniform nor monotonic. This discussion can be extended to multiple loci by considering larger contingency tables, more sophisticated multivariate discrete probability distributions and also multiple coefficients of association. Our data once again consists of a large number of discretized observations with a possibly complex correlation structure between the observations; suggesting the use of a graph theoretical formulation. If we represent each locus by a vertex, the LD structure can be captured graphically by introducing an undirected edge between a pair of vertices whenever the LD between the vertices in the pair is statistically significantly different from zero. It is assumed that there is a user defined significance threshold. The edge is undirected because the statment of statistical association between loci relates only to correlation, and does not carry any implications of causality. This defines an undirected graph whose edge structure is indicative of the LD patterns between the loci under consideration. Unlike in (Fishelson & Geiger, 2002); (Fishelson & Geiger, 2004); we do not associate states with a vertex in a graph, all that information has been averaged over all indviduals in determining the LD between markers. The use of graphical models to elucidate the LD structure between loci is well established (Thomas & Camp, 2004) as is the connection between graphical models and discrete multivariate probability distributions (Lauritzen, 1996). To summarize, datasets in statistical genetics consist of a vast collection of discretized observations with potentially very complex correlations between the observations; graph theoretical methods can be adopted for describing and analyzing data of this nature. In the rest of this chapter, we will discuss some applications of this nature, some open problems and possible solutions.

2. Graphical methods in association mapping

2.1 Population stratification

Understanding the LD structure is of considerable interest not only from the viewpoint of population genetics, but is vital for deducing the location of genes influencing traits in populations by Genetic Association Mapping. The Case Control design is a popular design for Genetic Association Mapping (Thomas, 2004). Here the data are assumed to consist of a large number of genotypes at fixed loci observed on two distinct groups of individuals, healthy controls and diseased cases, and we assume that the individuals are unrelated to each other. We assume that the genotypes are observed at marker loci, *i.e.* locations on the chromosome where there are no genes directly influencing the disease. Nonethless, if genotypes are observed at a large number of sufficiently closely spaced markers , there may be some markers physically close to the gene influencing the disease and which are potentially in strong LD with the diesease gene leading to a statisticaly significant association between certain genotypes and disease status. The goal of Case Control studies is to discover which markers (if any) show statistically significant associations with disease status and then draw conclusions about the physical location of a gene causing the disease with relation to these markers. As the association is statistical in nature it is important to understand potential causes of false positives in order to minimize Type I error. Two very important causes of Type I error in Case Control studies are population stratification and multiple testing artifacts. As we will see, undirected graphical models can be used to acquire new insights on both these problems. We begin with an analysis of population stratification; population stratification can be understood in terms of a difference in genetic content between cases and controls over and above any differences at loci in high LD with the disease gene. For example, if all the diseased cases are from one ethnic group, and all the healthy controls are from another ethnic group, then there will be statistically significant associations between case/control status and genotypes at loci which reflect differences in ethnicity *i.e.* population structure , in addition to loci which are possibly liked to the disease. This is an example of population stratification, and will lead to false positive associations at loci reflecting the ethnic differences between cases and controls but unrelated to the disease under study. In this very simple instance we just considered two ethnic groups, with all the cases drawn from one group and all the controls from the other, in more typical instances, both the cases and controls will themselves be mixtures of two or more ethnic groups. In these more realistic scenarios population stratification will be a problem when the proportions of various distinct groups are different in cases and controls and when genotypes are observed at loci where the frequencies of the various genotypes are different in the various ethnic groups represented in the case and control samples. The effects of population stratification can be ameliorated by carefully matching ethnicities between cases and controls but this is not always possible or feasible. In real life situations, it is safer to assume that population stratification exists, which then must be taken into account before testing any markers for association with disease status. There are two broad approaches to correcting for population stratification in genetic association studies, non-parametric and parametric. The most popular parametric method for dealing with population structure is using the program *Structure* (Pritchard, 2000) ;(Falush, 2003) in which specific scenarios for population admixture are assumed. *Structure* attempts to assign the individuals to specific clusters based on a specific model, the genotypes are the feature vectors used to decide how to assign individuals to clusters. In addition to the genotype data the number of populations present in the data must be supplied by the user, this is analogous to specifying the number of clusters in k-means or other clustering methods. In the most sophisticated scenario (the Linkage Model) the genotypes for any individual reflect a mixture of different populations with different chromosomal segments possibly arising from different populations. The number of populations however is not specified and must be supplied by the user. The precise assignment of individuals to different populations frequently arises only after a long MCMC simulation which uses genotypes for all individuals at all loci as input. Any LD between the loci is corrected for in the process of assigning individuals to constituent sub-populations. Since it is not uncommon to have genotypes at tens of thousands (frequently more) of loci implementing the methodology of *Structure* can be time consuming partly due to the sheer size of the data set and partly due the overhead of correcting for LD between the loci which is typically present when there are a large number of loci under consideration. The presence of LD between the loci also suggests that even though the number of loci may be large, the various loci do not necessarily contribute additional independent information on population stratification. This suggests that with a judicious choice of mutually independent loci, population stratification can be analyzed with a smaller and more manageable subset of the data in less CPU time. We will next explain how exactly this can be done using graph theoretical ideas and mention an application to real data.

An optimal set of loci for discerning population structure should be sufficiently large so that loci characteristic of populations whose frequency in the sample is relatively small, are nonetheless included, while ensuring a high degree of statistical independence between the loci. The requirement of statistical independence between the loci can be made more precise by ensuring that the loci are in low LD with one another. What we are then looking for is the largest possible subset of loci such that the LD between any arbitrary pair of loci is low. We will recast the problem in the language of graph theory using the correspondence between vertices in a graph and loci on a chromosome we discussed earlier and show a correspondence between a well known combinatorial optimization problem, that of finding the maximum independent set on an undirected graph. As there is no known polynomial time solution for this problem, a randomized heuristic algorithm will be described along with its performance on a real data set.

The input to the algorithm is $\mathcal{N}$, a set of N markers, an $N \times N$ symmetric matrix $\mathcal{M}$ with positive off diagonal values, and a positive constant c. The precise value of the diagonal matrix elements of $\mathcal{M}$ are not relevant as long as they are smaller than c. If we denote the elements of $\mathcal{N}$ by N_j ($1 \leq j \leq N$) then each row of $\mathcal{M}$ corresponds to a unique marker; with this assignment $\mathcal{M}_{ij}(i \neq j)$ is simply the magnitude of the association between markers N_i and N_j. All the different M_{ij} can be easily computed given a rectangular data matrix of genotypes in which individuals are indexed by rows and each column contains the genotypes for one particular marker. As mentioned earlier we assign to each marker a vertex in an undirected graph $\mathcal{G}$. Thus $\mathcal{G}$ has a set of vertices $\mathcal{V}$ with N elements denoted by $V_i \in \mathcal{V}$, where $1 \leq i \leq N$. Since each marker is assigned to a unique row of $\mathcal{M}$, we can now uniquely associate to each row of $\mathcal{M}$ a vertex $V_i \in \mathcal{V}$, where $1 \leq i \leq N$. Let the set of edges of $\mathcal{G}$ be denoted by $\mathcal{E}$. An undirected edge E_{ij} exists between any two elements V_i and V_j of $\mathcal{V}$ if $\mathcal{M}_{ij} > c$. This condition is adequate to define all the elements of $\mathcal{E}$. There can clearly be no edges from any vertex to itself due to the choice of the diagonal matrix elements of $\mathcal{M}$. By a suitable choice of c, any two unlinked vertices in $\mathcal{G}$ can be made to correspond to two unassociated markers in $\mathcal{N}$. The precise value of c needed to achieve this correspondence will depend on some user specified threshold for defining significant association. Thus given a subset of vertices $\{V_i, V_j, V_k, V_m\}$ with no edges connecting any of the six possible pairs of vertices that can be formed from this subset, we can find a corresponding subset of markers $\{N_i, N_j, N_k, N_m\}$ which are mutually unassociated. This argument can be extended to any $\mathcal{V}_s \subset \mathcal{V}$ which gives rise to a corresponding $\mathcal{N}_s \subset \mathcal{N}$ of mutually unassociated markers. Furthermore, each unique $\mathcal{N}_s \subset \mathcal{N}$ corresponds to a unique $\mathcal{V}_s \subset \mathcal{V}$. However, any $\mathcal{V}_s$ corresponds to a clique on $\mathcal{G}^c$, the complement graph of $\mathcal{G}$. If we want the largest possible $\mathcal{N}_s \subset \mathcal{N}$ of mutually unassociated markers we must find the maximum independent set of vertices in $\mathcal{G}$.

We have thus transformed the problem of finding the largest possible set of mutually unassociated markers to a well known problem in graph theory that of finding the maximum independent subset of vertices in an undirected graph, (or equivalently the clique of largest size on the complement graph) a problem for which there is no known efficient solution. Thus we are forced to resort to heuristics which yield only approximate solutions, more precisely a subset of vertices which may be smaller in size than the true maximum independent subset. As a cross-check on any given solution it would be useful to have a different solution for the sake of comparision. This motivates the use of a stochastic greedy heuristic which can generate multiple solutions, rather than use of one of the many well known published heuristic algorithms for this problem. The graph that we have is unweighted, although we could have considered a weighted graph with the LD between markers playing the role of weights. Although the precise LD information has been ignored in the construction of the graph and in our heuristic algorithm, this does not matter. LD represents a statistical correlation and all that matters for our purposes is whether the correlation is significant or not. One complication we have ignored here is that the presence or absence of edges is determined by comparing sample LD values with some threshold; in a more sophisticated scheme the standard errors on the sample LD values could also be used to assign probabilities for the presence of edges in the graph where the corresponding sample LD values are close to the significance threshold.

We will next describe a stochastic greedy heuristic for finding the clique of maximum size on an undirected graph, which due to the exact corespondence between maximum cliques and maximum independent sets can readily be applied to our maximum independent set problem.

• Description of Algorithm

As before we assume we have an undirected graph $\mathcal{G}$ in which $\mathcal{V}$ is the set of vertices and $\mathcal{E}$ is the set of edges. $\mathcal{G}$ is not assumed related to any genetic marker map so the algorithm is at this stage perfectly general. The algorithm also requires as input a positive parameter γ which is a measure of how far the algorithm deviates from a deterministic greedy heuristic. The larger the value of γ the closer the algorithm resembles a deterministic heuristic. We define L_i the set of neighbors of V_i and n_i size of L_i, and assume $n_i > 0 \ \forall \ V_i \in \mathcal{V}$. We define sets of vertices *CandSet*, *TempSet* as well as *ReturnSet* which is the output from the program.

Informally, the algorithms starts by picking a seed vertex which has a relatively large number of neighbors, relatively large being defined with respect to the number of neighbors of all the vertices in the graph. This seed vertex V_s is inserted into *ReturnSet*. *CandSet* is initialized by the set of neighbors V_s, while *TempSet* is initialized by the empty set. An element V_n of *CandSet* is chosen on the basis of having a relatively large number of neighbors and *TempSet* is the set of neighbors of V_n not already included in *ReturnSet*. Next *CandSet* $\leftarrow$ (*CandSet* $\cap$ *TempSet*), which has the effect of ensuring that all all surviving elements of *CandSet* are elements of both V_s and V_n, i.e. all elements of *CandSet* are connected to all elements of *ReturnSet*. Once this step is carried out, it is safe to pick another element from *CandSet* and repeat the cycle untill *CandSet* is the null set. At this stage *ReturnSet* cannot be further augmented and the algorithm halts. A more precise description of the algorithm is given below.

• Initialization

1. Compute $norm = \sum\limits_{i=1}^{N} n_i^{\gamma}$.
2. Evaluate $p_i = (n_i^{\gamma}/norm) \ \forall \ i \ 1 \leq i \leq N$.
3. Pick some j with probability p_j.
4. Insert V_j in *ReturnSet*.
5. *CandSet* $\leftarrow L_j$.
6. *TempSet* $\leftarrow \emptyset$.

• Main Loop

while *CandSet* $\neq \emptyset$ do

    1. Evaluate $n_i^{\gamma} \ \forall \ V_i \in CandSet$
    2. Compute $norm = \sum n_i^{\gamma}$
    with the summation restricted to elements of *CandSet*
    3. Compute $p_k = (n_k^{\gamma}/norm) \ \forall \ V_k \in CandSet$
    4. Select $V_n \in CandSet$ with probability p_n.
    5. Insert V_n in *ReturnSet*.
    6. *TempSet* $\leftarrow \{V_m \in L_n : V_m \notin ReturnSet \text{ where } 1 \leq m \leq N\}$
    7. *CandSet* $\leftarrow$ (*CandSet* $\cap$ *TempSet*)
    8. *TempSet* $\leftarrow \emptyset$

If *CandSet* $\equiv \emptyset$ return *ReturnSet*.

It is also possible to define a deterministic greedy heuristic in which the vertices selected in step 3 of the initialization and step 4 of the main loop are just those with the largest number of neighbors. If the parameter γ is made larger and larger, then the output from the program will increasingly resemble that from a deterministic greedy heuristic. It is worth pointing out that our algorithm does not make use of the relative location of markers with respect to each other along the chromosome, thus the methodology we outline is applicable whether the LD decays rapidly or slowly as a function of the distance between markers on the chromosome. The output from the algorithm returns a subset of markers which is not necessarily the largest subset of independent markers, nontheless the number of markers returned could still be large enough to get an accurate handle on population stratification. We now briefly discuss the application of of this to real data, more details are available in (Hamblin, 2010).

In conjunction with the Barley Coordinated Agricultural Project (www.BarleyCAP.org), 1816 Barley lines (treated as individuals for our purposes) were genotyped at 1415 markers. Five initial attempts to run *Structure* with 500,000 iterations and 100,000 burn in steps taking into account the association between markers and allowing for admixture between populations were unsuccessfull due to non-convergence of the MCMC iterations. At this stage, our algorithm was used to identify a subset of markers with linkage disequilibrium r^2 between any two markers in the subset to be less than 0.25, a criteria used to decide when to consider markers unlinked. A subset of 486 markers was identified and used as input for *Structure* allowing for admixture between individuals but no association between markers. Eight runs of *Structure* with 100,000 burn in and 200,000 analysis iterations all converged with consistent likelihood estimates, illustrating the utility of selecting unassociated markers as opposed to using the entire set and allowing for association between markers. It is worth pointing out that constituent populations identified by *Structure* have differing linkage disequilibrium structure at both short and large distances, with some SNPs in a few but not all of the subpopulations in high LD even when 50cM apart. As mentioned earlier, our algorithm does not use map distances in selecting markers, neither the presence of significant LD between unlinked markers nor the very different patterns of LD in the subpopulations is an issue. This feature gives rise to a complex edge structure on the graph, similar to the examples considered in (Thomas & Camp, 2004).

We next turn our attention to the relevance of the maximum independent set problem to non-parametric methods for analyzing population stratification. The most widely used non-parametric approach for analyzing population stratification is Principal Components Analysis, a number of popular implementations such as EIGENSTRAT (Price, 2006) and EIGENSOFT (Patterson, 2006) are available. We will discuss the relevance of the apparoaches just described to EIGENSOFT and then show how some of the statistical methodology in EIGENSOFT might have applications to machine learning problems outside of statistical genetics. The input data for EIGENSOFT is a rectangular data matrix where the rows correspond to individuals and there is one column for each marker, the entries of the data-matrix correspond to the genotypes suitably parametrized and standardized. The key idea behind the implementation in (Patterson, 2006) is the realization that in the absence of population stratification, the largest Singular Value of the data matrix is distributed according to the Tracy Widom distribution (Tracy & Widom, 1994). However, even in the absence of stratification, deviations from the Tracy Widom distribution are possible if there is LD between the markers. One way to avoid false signals of population stratification is to choose a set of markers which are mutually uncorrelated, preferrably as large an unrelated set of markers as possible. As we have seen in the discussion of model dependent population stratification, choosing this set is tantamount to solving an instance of a maximum independent set on an undirected graph defined by the LD matrix. In practical instances, alternative methods have been used to find a set of unrelated markers, for example by exploiting special features of the LD structure or by other approximations(Heerwarden, 2010). It is not clear that these approaches will find the largest possible set of uncorrelated markers which is required for putting the most stringent bounds on the extent of population stratification. Very few (if any) attempts have been made to use the methodology previously described to identify as large a subset of unrelated markers as possible, such an analysis could be fruitful. This concludes our discussion on the application of graph theoretical methods for analyzing population stratification. The key point of our discussion is the relevance of the problem of finding the maximum independent set to understanding population stratification. This connection has not been established before (to the best of our knowledge) and can be exploited to speed up the the analysis of population stratification in real data sets. Before going on to discuss the application of graph theoretical methods to multiple testing, it is worth mentioning how the methodology developed in EIGENSOFT could possibly be used to address a long standing problem in cluster analysis, *i.e.* how to identify the number of distinct groupings in a dataset. As mentioned in our discussion of *Structure* the number of constituent populations to fit in *Structure* is user defined, in (Patterson, 2006) the authors point out how this number might be reliably estimated using the sample singular values of the data matrix and the details of the Tracy Widom Distribution. What this amounts to is computing a non-parametric statistic of the dataset which is then used to estimate the number of distinct groupings in the dataset. Since the methodology is very general and model independent it could conceivably be applied to a whole range of problems far removed from statistical genetics.

2.2 Multiple testing

Another major source of Type-I error in Genome Wide Association studies (GWAS) is false positives arising from multiple testing, and as mentioned earlier these can arise even if population stratification between cases and controls is fortuitously negligible or has been controlled for in some manner. Before we discuss the relevance of graph theoretical methods for understanding multiple testing artefacts, it is worth outlining the root of the problem and some common remedies. A large number of markers (N) are tested one after another for association with the trait or disease of interest. Under H_0 none of the markers are associated with the trait, and in addition the p-values for the test statistic are distributed like $\sim U(0,1)$. For a significance level α the expected number of significant tests under H_0 will be $\sim N\alpha$, since N in modern GWAS can be $\mathcal{O}(10^6)$ this leads to a sizeable number of false positives even if α is small. One way to ensure that there are no false positives with N independent tests is by choosing α so that $N\alpha \ll 1$ (the Bonferroni correction). However this leads to such stringent significance thresholds that only markers with very strong effects are picked up and many markers associated with the trait are ignored because their effects are not large enough to survive the stringent significance threshold, *i.e* there is is a large Type II error rate. Furthermore, if all the tests are not independent due to correlations between the markers the correction is excessively conservative. This problem can be avoided by permutation testing which leads to a non-parametric estimate of the number of significant test under H_0 given the correlation structure between the markers. While this approach certainly works it can become computationally very intensive when there are hundreds of thousands of markers to be tested. A less computationally intensive method to lower the number of Type II errors at the cost of allowing a certain number of false positives is by controlling the False Discovery Rate (FDR)(Benjamini & Hochberg, 1995); variants on this idea have also been considered. It has been suggested by (Nyholt, 2004) that the Bonferroni corrections be modified by replacing N with N^*, the number of independent tests, where hopefully N^* is very much smaller than N. With this replacement, the significance threshold can be made less stringent lowering the Type II error rate. We will next examine this sugestion in the language of graph theory, more specifically we will consider the problem of finding a suitable subset of independent markers, the size of this subset is the number of independent markers. One key observation is that restricting ourselves to a subset of markers is most meaningfull if it is possible to choose that subset of markers not only to be statistically independent also to be serve as surrogates for all the markers under consideration. If the latter condition is fulfilled, then testing only the markers in the independent subset can be regarded as testing each and every marker. If this condition is not fulfilled, we run the risk of skipping association tests on some markers which are part of the panel. It is worth pointing out that the idea that a limited subset of markers can be used as surrogates for an entire panel is well established; this is the notion underlying the use of tag SNPs in GWAS (Carlson, 2004). Identifying an optimal tag SNPs in a marker panel given the LD matrix between the markers can be reformulated as a variety of different well known graph theoretical problem, including the search for a dominating set of smallest size (Li & Wang, 2011). In order not to underestimate the number of independent tests it is necessary that the subset of markers be as large as possible; from our earlier discussion of population stratification it is clear than once again we are dealing with finding a maximum independent set on a undirected graph defined by the LD structure between the markers and a user defined specification of statistical independence between markers. The condition that the markers we select be proxies for all the markers in the panel can be fulfilled by requiring that each vertex be connected by an edge to at least one of the markers in the maximum independent set. In other words, the maximum independent set should be a dominating set for the graph. Since any maximal independent set is a dominating set (Foulds, 1992) the maximum independent set satisfies this condition. If the heuristic used to find the maximum independent set only returns a maximal independent set, this attractive feature will be preserved. Thus estimating the number of independent tests via maximum independent set heuristics seems to have some advantages. One can also approach the problem of estimating N^* in terms of the the size of the smallest dominating set on the graph. If the smallest dominating set turns out not to be an independent set, then the resulting estimate of N^* would be smaller than what we would obtain from analyzing independent sets, but not easy to estimate precisely, given the dependence of the markers. This point is illustrated in Fig. 1. where $\{A, B, E, F\}$ is the maximum independent set, but all markers can be tested by considering just two (not independent) markers C and D. In situations such as these, it is not clear what to value use for N^*.

Fig. 1. Ambuguity in N^*

In practise implementing the prescription of (Nyholt, 2004) has been shown to be problematic in real and simulated datasets (Dudbridge & Koeleman, 2004);(Salyakina, 2005);(Coneely & Boehnke, 2007), but there has been no general model independent analysis as to why these difficulties arise. Our graph theoretical analysis sheds light on possible ambiguities in the prescription of (Nyholt, 2004) which may be at part of the reason for its observed limitations. Before concluding our discussion on the applications of the maximum independent set it worth recalling that all that heuristics can deliver is a lower bound on the size of the maximum complete set. What is missing is some means using the data to obtain an estimate of the upper bound, such an estimate could potentially improve the performance and applicability of the heuristics. One feature of the data which has not been exploited is that the $N \times N$ matrix of correlation values is obtained from a data matrix with dimensionality $N_{ind} \times N$, where N_{ind} is the number of individuals and typically $N_{ind} < N$. Thus both the data matrix and the correlation matrix can be expected to have rank less than N. The redundancy in the rows of the correlation matrix due to the reduced rank could possibly be exploited in order to obtain an upper bound on the size of the maximum independent set. As has been shown in (Li & Li, 2005) this redundancy can be used to obtain an alternative estimate for the number of independent tests; combining the approach of (Li & Li, 2005) with the maximum independent set heuristic we describe here could be a fruitfull line of future research.

3. Blocking Gibbs

In the remainder of this chapter we will focus on possible applications of graph theoretical methodology to analysing pedigree data; more precisely we will consider individuals with known parent offspring relations and genotypic information at a possibly large number of loci. As was mentioned earlier, the pedigree structure and the genotypes can be combined to form a Bayesian Network where the conditional probabilities along the edges are defined by Mendelian Genetics. Since there are known genotypes there are vertices in the Bayesian networks where evidence is available. From the standpoint of genetic linkage analysis one of the most important quantities to be computed from a pedigree and associated genotypic data is the Likelihood (Ott, 1999). Computing the Likelihood involves evaluating a very complex series of nested sums and products of conditional probabilities over expressions such as the one shown below:

$$\cdots P(g^G \mid g^F, g^E) \, P(g^H \mid g^F, g^E) \, P(g^E \mid g^A, g^B) P(g^I \mid g^C, g^D) P(g^C \mid g^A, g^B) \, P(g^D \mid g^A, g^B) \cdots \tag{1}$$

g^A, g^B, etc. are discrete random variables representing either genotypes or alleles, and the $\cdots$ indicate the presence of many more such conditional probabilities. The conditional probabilities shown above are typical of the factors that would appear in the Joint Probability Distribution defined by the Bayesian Network, however realistic pedigrees often contain far more factors than can be written down. Computing the Likelihood involves summing over all allowed values of all the random variables, (i.e. all consistent genotypes), and in realistic situations where there are huge numbers of conditional probabilities, this is analytically intractable. Numerical solutions are hypothetically possible due to the local structure of the computations (Lauritzen & Spiegelhalter, 1988). The computational effort involved depends critically on the order in which the summations are performed (Jordan, 2004) and determining the lowest cost summation order is $\mathcal{NP}$ Hard (Arnborg, 1987). If no good heuristic algorithm for determining the most efficient summation order can be found the multiple sum cannot be performed exactly, and must be approximated by sampling the most significant terms.

This provides the motivation for introducing Markov Chain Monte Carlo (MCMC) methods which have been used extensively in linkage analysis (Thompson, 2005). The simplest form of MCMC sampler to implement is the Gibbs sampler, which however can be very tricky to implement on a large pedigree with many known genotypes. One of the key conceptual difficulties in implementing the Gibbs Sampler can be illustrated on a very simple situation involving just four individuals as shown in the following figure.

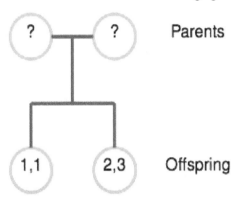

Fig. 2. Trouble with Gibbs

The two offspring have observed genotypes, and the laws of Mendelian Inheritance dictate that the two parental genotypic combinations are either $\{\{1,2\},\{1,3\}\}$ or $\{\{1,3\},\{1,2\}\}$ in obvious notation. In order for the sampler to be irreducible transitions between configurations should be possible, and in keeping with textbook Gibbs sampling one parental genotype would be updated at a time keeping the other fixed. Let us begin from the configuration $\{\{1,2\},\{1,3\}\}$ with a view to reaching $\{\{1,3\},\{1,2\}\}$. If we sample conditional on any one parent and the known genotypes, there is no way in which we can update the genotype of the other parent; *i.e.* the sampler gets stuck in the starting configuration. Thus a single site update is problematic and it is easy to see that the root of the problem lies in the stringent constraints arising from Mendelian Genetics which lead to strong correlations between the variables to be updated. A possible solution within the framework of Gibbs sampling is to update both parental genotypes simultaneously, *i.e.* use a blocking Gibbs sampler where a block consists of multiple stochastic variables which are strongly correlated and must be updated simultaneously. The idea of updating multiple strongly correlated variables during a single MCMC update in order to improve convergence is well established and outperforms standard Gibbs sampling in statistical genetics (Totir, 2003) and other applications. (Swendsen & Wang, 1987); (Roberts & Sahu, 1997). Furthermore, this approach has been applied in Bayesian Networks arising not only in Statistical Genetics (Jensen & Kong, 1999);(Thomas, 2000), but also in expert systems (Jensen, 1995). For our purposes the optimal choice of blocks is not only crucial for ensuring the irreducibility of the sampler but also for improving the mixing and convergence properties of the sampler. In the rest of this chapter we will study the issue of block definition and relate this problem to a well known problem in machine learning, that of partitioning data sets into semi-autonomous clusters. Before doing so we will briefly mention another aspect of likelihood computations on large pedigrees which has attracted recent attention, *i.e.* the relation to constraint satisfaction. The problem of finding assignments of unknown genotypes consistent with known genotypes, the pedigree structure and the laws of Mendelian Inheritance can be viewed as finding the solution of a constraint satisfaction problem, and is known to be computationally hard (Aceto, 2001). There is however one additional complication which arises in dealing with pedigrees, each solution can be assigned a posterior probability and what is required are the solutions with higher posterior probabilities. What an irreducible ergodic MCMC sampler should do is not only find solutions to a very complex contraint satisfaction problem, but also assign the correct posterior probability to the various solutions. Seen in this light it is easy to see why the MCMC sampling on pedigrees can be so challenging.

The key difficulty in constructing blocks is correctly grouping strongly correlated variables together, followed by updating them simultaneously in a manner consistent with the known genotypes. In simple instances like Fig. 2, grouping variables is easy, but in more complicated cases such as the pedigree in Fig. 3 it can be highly non-trivial.

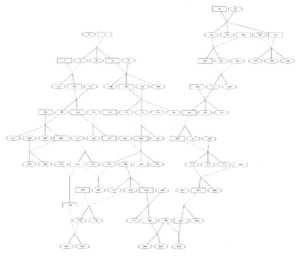

Fig. 3. Large pedigree

A moments reflection will suggest that the difficulty in partitioning the pedigree of Fig. 3 into blocks arises because of the large number of cycles in the graph. A more general analysis of the difficulties has been undertaken in (Jensen & Sheehan, 1998), the presence of cycles is indeed a problem and no general solution for dividing an arbitrary pedigree into blocks is known. However, a few features of a well motivated Blocking Scheme can be identified

- The assignment of blocks should be such that all the variables are assigned to at least one block. If this condition is not satisfied, some variables may not be updated leading to biased MCMC estimates.
- All strongly correlated variables must be contained in the same block, if not the irreducibility issues mentioned earlier will arise
- Within a given block variables should be more strongly correlated with each other than with variables outside the block.

It is easy to understand the relevance of these criteria, the last two criteria are not only relevant for pedigree analysis but are also similar to what might expected from an optimal partitioning of a data set into clusters. Note that we have not specified the number of partitions in advance; if this scheme were to be implemented on an arbitrary dataset it would not only assign all the elements of the dataset to clusters but could also return the number of partitions. Thus any heuristic solution for finding optimal blocks could be very useful in a number of machine learning applications. One important distinction between pedigree analysis and other applications is the requirement that the sampler mix rapidly, this was the motivation for the use of overlapping blocks. The relevance to data partitioning problems of a more general nature is greatest when the clusters are expected to overlap. The other important distinction is that many genotypes in pedigrees may be unknown, corresponding to vertices with no information. Missing or ambiguous data are not as widely considered in other data sets, so the analogy works better when there are not too many unknown genotypes.

One outline of a Blocking Gibbs scheme was made in (Abraham, 2007) where the problem of Block Identification and consistent assignment of genotypes were addressed simultaneously. The algorithmic insight exploited in (Abraham, 2007) was that the genotypes of any given individual are strongly dependent by just a handfull of close relatives; in the language of the pedigree graph the state of a vertex is influenced by just a handfull of neighbouring vertices. This is because the edge structure in the graph reflects a combination of either relatedness between individuals or physical distance between loci. The notion of neighbouring vertices can be made more precise by defining distances between vertices in terms of a breadth first search. Due to the underlying Markov Field Property, vertices which are far apart as defined by the breadth first search are expected to be roughly independent. Once there is a guideline for deciding which vertices can be expected to be independent of other vertices, it becomes possible to partition the graph into overlapping blocks in which consistent genotype assignments in a block can be made with little input from the evidence from other blocks. The dataset used in (Abraham, 2007) is very complex and has many of the features discussed in (Jensen & Sheehan, 1998) which are known to lead to difficulties, nonetheless it was possible to generate a consistent set of genotypes using the scheme just outlined. Furthermore, it was checked that the posterior probabilities of the genotypes found in this manner were consistent with those that would have been obtained in the absence of any approximations. This suggests that separation of vertices on the graph is a useful guideline for assessing the approximate independence of the corresponding random variables. Criteria similar to these have been successfully used to construct blocks and mcmc samplers in other complex examples, (Habier, 2009);(Habier, 2010) indicating that the basic idea may have a broad general applicability.

If we consider the problem in a more general light, what we have done is to use the known correlations in a data set containing many discrete observations to identify subsets of variables which have strong correlations with each other but weaker correlations with the other variables. If this methodology were to be applied to cluster a general data set with a a known matrix of correlation values it would be first necesary to define graph and identify edges between the vertices (datapoints). Identifying edges could be acchieved through a user defined threshold which could be defined independent of the data values as described in our discussion of population stratification , or could be defined in terms of some suitable number of sample standard deviations above the sample mean of all the correlation values. Once the edges are specified in this manner, the procedure used in (Abraham, 2007) could be used to define blocks which in a more general case would correspond to a cluster in the data set. One advantage of this procedure is that it has been shown to work in the context of pedigree graphs where inaccurate assignment of vertices to clusters will often be penalized by poor

MCMC convergence or in extreme cases by a lack of irreducibility of the sampler. Adapting the blocking methodologies in (Abraham, 2007);(Habier, 2009) and (Habier, 2010) to other cluster identification in general ohmics data sets could prove to be fruitful.

We next consider a long standing issue which is relevant in both block assignment in blocking gibbs and cluster assignment in general, *i.e.* the problem of determining the number of independent subgroups in the data set making as few model dependent assuptions as possible. As was mentioned in our discussion of non-parametric population stratification , the authors of (Patterson, 2006) suggest that from the elements of a suitably constructed correlation matrix a test statistic can be obtained which can be used to decide on the appropriate number of populations to use as input for parametric population stratification analysis. The treatment of this issue in (Patterson, 2006) is so general that it would appear to be the basis for a model-free approach that could be used to estimate the number of subgroups in an arbitrary ohmics data set given a matrix of correlation values. As applied to blocking gibbs, the matrix of correlation values could be substituted by the distance matrix used in (Abraham, 2007) or some more sophisticated variant thereof. In this regard it is worth recalling that number zero eigenvalues of the Laplacian of an undirected graph is the number of connected components, which supplies a lower bound on the number of clusters. Thus the connection between the entries of a suitable correlation matrix and the number of clusters is well established, by applying the results of (Patterson, 2006) it might be possible to extract more detailed information on the number of clusters present in a dataset.

4. Conclusions

In this chapter we have discussed the relevance and applications of graph theoretical methods to a number of problems in statistical genetics. In particular, some novel applications of the maximum independent set on an undirected graph to population stratification were presented. Some key issues in the construction of Blocking Gibbs Samplers on complex pedigrees were discussed along with their relevance outside of statistical genetics.

5. Acknowledgments

RLF and KJA both received supprot from the United States Department of Agriculture, National Research Iniative Grant USDA NRI-2009-03924. KJA also acknowledges financial support of the program Professor Visitante do Exterior of Coordenação de Aperfeiçoamento de Pessoal de Nível Superior (CAPES) , Brasil. KJA thanks Prof. Jean-Eudes Dazard and members of the Department of Computation and Mathematics, University of São Paulo (Ribeirão Preto) for valuable discussions.

6. References

Abraham, K.J., *et.al.*. (2007). Improved techniques for sampling complex pedigrees with the gibbs sampler, *Genetics, Selection and Evolution* 39: 27–38.

Aceto, L., *et. al.*. (2001). The complexity of checking consistency of pedigree information and related problems, *The Journal of Computer Science and Technology* 19(1): 42–59.

Arnborg, S., *et.al.*. (1987). Complexity of finding embeddings in a k-tree, *SIAM Journal on Algebraic and Discrete Methods* 8: 277–284.

Benjamini, Y. & Hochberg, Y. (1995). Controlling the false discovery rate a practical and powerful approach to multiple testing, *Journal of the Royal Statistical Society B* 57: 289–300.

Carlson, C.S., *et.al.*. (2004). Selecting a maximally informative set of single-nucleotide polymorphisms for association analysis using linkage disequilibrium, *American Journal of Human Genetics* 74: 106–120.

Coneely, K. & Boehnke, M. (2007). So many correlated tests, so little time! rapid adjustments of p values for multiple correlated tests, *American Journal of Human Genetics* 81(6): 2074–2093.

Dudbridge, F. & Koeleman, B. (2004). Efficient computation of significance levels for multiple associations in large studies of correlated data, including genomewide association studies, *American Journal of Human Genetics* 75(3): 424–435.

Falush, D., *et.al.*. (2003). Inference of population structure using multilocus genotype data: Linked loci and correlated allele frequencies, *Genetics* 164: 1567–1587.

Fishelson, M. & Geiger, D. (2002). Exact genetic linkage computations for general pedigrees, *Bioinformatics* 18 Suppl. 1: S189–S198.

Fishelson, M. & Geiger, D. (2004). Optimizing exact genetic linkage computations, *Journal of Computational Biology* 11(2-3): 263–275.

Foulds, L. (1992). *Graph Theory Applications*, Springer Verlag.

Habier, D., *et.al.*. (2009). Genomic selection using low density marker panels, *Genetics* 182: 343–353.

Habier, D., *et.al.*. (2010). A two-stage approximation for analysis of mixture genetic models in large pedigrees, *Genetics* 185: 655–670.

Hamblin, M.T., *et.al.*. (2010). Population structure and linkage disequilibrium in us barley germplasm: Implications for association mapping, *Crop Science* 50(2): 556–566.

Heerwarden, J. V., *et.al.*. (2010). Fine scale genetic structure in the wild ancestor of maize zea mays ssp parviglumis, *Molecular Ecology* 19: 1162–1163.

Jensen, C. & Kong, A. (1999). Blocking gibbs sampling for linkage analysis in large pedigrees, *American Journal of Human Genetics* 65(3): 885–901.

Jensen, C. & Sheehan, N. (1998). Problem with determination of noncommunicating classes for markov chain monte carlo applications in pedigree analysis, *Biometrics* 54: 416–425.

Jensen, C.S., *et.al.*. (1995). Blocking gibbs sampling in very large probabilistic expert systems, *International journal of human computer studies* 42(6): 573–704.

Jordan, M. (2004). Graphical models, *Statistical Science* 19(1): 140–155.

Lauritzen, S. L. (1996). *Graphical Models*, Oxford University Press.

Lauritzen, S. & Spiegelhalter, D. (1988). Local computations with probabilities on graphical structures and their application to expert systems, *Journal of the Royal Statistical Society Series B* 50: 157–224.

Li, J. & Li, K. (2005). Adjusting multiple testing in multilocus analyses using the eigenvalues of a correlation matrix, *Human Heredity* 95: 221–227.

Li, J. & Wang, W.-B. (2011). Tag snp selection, *in* E. Zeggini & A. Morris (eds), *Analysis of Complex Disease Association Studies, A Practical Guide*, Academic Press, pp. 49–65.

Nyholt, D. (2004). A simple correction for multiple testing for single nucleotide polymorphisms in linkage disequilibrium with each other, *American Journal of Human Genetics* 74(2): 765–769.

Ott, J. (1999). *Statistical Methods in Genetic Epidemiology*, The Johns Hopkins University Press.

Patterson, N. et.al.. (2006).   Population structure and eigenanalysis, *PLoS Genetics* 2(12): 2074–2093.

Price, A.L., et.al.. (2006).   Principal components analysis corrects for stratification in genome-wide association studies, *Nature Genetics* 38: 904–909.

Pritchard, J.K., at.al.. (2000). Inference of population structure using multilocus genotype data, *Genetics* 155(2-3): 945–959.

Roberts, G. & Sahu, S. (1997).   Updating schemes, correlation structure, blocking and parameterization for the gibbs sampler, *Journal of the Royal Statistical Society Series B* 59(6): 573–704.

Salyakina, D., et.al.. (2005). Evaluation of nyholt's procedure for multiple testing correction, *Human Heredity* 60: 19–25.

Swendsen, R. & Wang, J.-S. (1987). Nonuniversal critical dynamics in monte carlo simulations, *Physical Review Letters* 58: 86–88.

Thomas, A. & Camp, N. (2004).   Graphical modelling of the joint distributions of alleles at associated loci, *American Journal of Human Genetics* 74: 1088–1101.

Thomas, A., et.al.. (2000). Multilocus linkage analysis by blocked gibbs sampling, *Statistics and Computing* 10: 259–269.

Thomas, D. T. (2004). *Statistical Methods in Genetic Epidemiology*, Oxford University Press.

Thompson, E. A. (2005). Mcmc in the analysis of genetic data on pedigrees, *in* W. S. Kendall, F. Liang & J.-S. Wang (eds), *Markov Chain Monte Carlo Innovations and Applications*, World Scientific Publishing, pp. 183–217.

Totir, L.R., et.al.. (2003). A comparison of alternative methods to compute conditional genotype probabilities for genetic evaluation with finite locus models, *Genetics,Selection and Evolution* 35: 585–604.

Tracy, C. & Widom, H. (1994). Level spacing distribution and the airy kernel, *Communications in Mathematical Physics* 159: 151–174.

Weir, B. S. (1996). *Genetic Data Analysis II*, Sinauer Associates, Sunderland MA 01375 USA.

3

Simulation of Flexible Multibody Systems Using Linear Graph Theory

Marc J. Richard
Department of Mechanical Engineering,
Laval University, Québec (Québec)
Canada

1. Introduction

This chapter provides a general description of a variational graph-theoretic formulation for simulation of flexible multibody systems (FMS) which includes a brief review of linear graph principles required to formulate this algorithm. The system is represented by a linear graph, in which nodes represent reference frames on flexible bodies, and edges represent components that connect these frames. To generate the equations of motion with elastic deformations, the flexible bodies are discretized using two types of finite elements. The first is a 2 node 3-D beam element based on *Mindlin* kinematics with quadratic rotation. This element is used to discretize unidirectional bodies such as links of flexible systems. The second, consists of a triangular thin shell element based on the discrete *Kirchhoff* criterion and can be used to discretize bidirectional bodies such as high speed lightweight manipulators, deployable space structures and micro-nano electro-mechanical systems (MEMS).

Realistic dynamic simulation of industrial mechanisms that requires tracking accuracy at high operational speed is becoming increasingly important for engineers. Hence, to accurately describe such motions, the effects of flexibility and damping must be included in the dynamic model. Since the equations governing the motion of FMS are highly non-linear and dynamically coupled, one must exploit some kind of linear graph principles (Andrews, 1971; Behzad & Chartrand, 1971; Christofides, 1975; Even, 1979; Koenig & Blackwell, 1960) to properly define the interconnection between the bodies. By combining linear graph theory (Andrews & Kesavan, 1975; Koenig et al., 1967; Richard, 1985) with the principle of virtual work (Richard et al., 2011; Shi & McPhee, 2000; Shi et al., 2001) and finite elements, a dynamic formulation is obtained that extends graph-theoretic (GT) modelling methods to the analysis of 3-D beams and shell surfaces of FMS. The widespread interest in flexible multibody systems (FMS) is evidenced by the existence of a large number of algorithms (Wasfy & Noor, 2003). It has been shown (McPhee & Redmond, 2006; Richard et al., 2004; Shi & McPhee, 1997) that for multibody systems, a graph-theoretic formulation can generate a minimal set of equations. GT-based approaches explicitly separate the linear topological equations for the entire system from the non-linear constitutive equations for individual components, resulting in very modular and efficient algorithms.

To be applicable to a wide range of spatial mechanical systems containing both open and closed kinematic chains, a flexible multibody formulation must incorporate general mathematical methods for representing both the system topology as well as the time-varying configuration. The representation of topology is naturally handled using elements of graph theory. It has also been shown (McPhee, 1998) that a proper "tree" with a set of coordinates called "branch coordinates" encompasses both sets of Cartesian and joint coordinates as special cases. Also, the use of virtual work has been proposed and validated as a new graph-theoretic variable. In order to create a system graph that results in correct kinematic and flexible dynamic equations for any choice of spanning tree, it is necessary to introduce a dependent virtual work element. Hence, by combining linear graph theory and the principle of virtual work, it is possible to develop a variational graph-theoretic formulation in terms of branch coordinates capable of automatically generating the motion equations for FMS.

2. System representation by linear graph

Many researchers have studied the theory of graphs (Arczewski, 1990; Baciu et al., 1990; Chou et al., 1986; Roberson, 1984) and bond-graphs (Bos, 1986; Hu, 1988) mainly due to the fact that among all the fields of human interest, there are few where graph theory cannot be applied to the process of analyzing or synthesizing problems. In order to extract the kinetic properties resting within mechanical systems, it is convenient to discretize the system into a schematic diagram composed of nodes or vertices representing points of interconnection in the system and oriented edges identifying system elements. Combination of all the vertices delimiting the network of elements with the total set of spanning edges between appropriate nodes will result in a diagram which is a simple isomorphism of the mechanical system.

One of the most appealing features of graph-theoretic methods lies in the geometric and pictorial aspect of the method. Given a spatial mechanical system, one can construct the system's diagram by a simple mapping of the mechanism. For instance, the vertices would correspond to rigid or flexible bodies, points on bodies to which forces are applied or joints are connected, and a ground node that represents the origin of an inertial reference frame. Each element is represented by a line segment and each joint or connection by an appropriate point such that a user can associate the network diagram to the mechanical system in a direct fashion. The technique is very methodical and well suited for computer implementation.

Much of the simplicity and efficiency of graphical methods lie in the use of a "tree" to assist in arranging the order of computation. By definition, a tree is a subgraph where every vertex of the graph is connected by exactly one chord. This connotes that the subgraph is connected and contains all the nodes of a given system graph, but has no closed loops. A tree is considered a minimal connected graph in which the deletion of a single branch would separate the subgraph. The set of vectors that complement the tree are called "cotree". It has been shown (McPhee, 1998) that the components necessary to generate an optimum tree for branch coordinates should be selected in the following order: N_1 rigid or flexible arms; N_2 position drivers (function of time); N_3 spherical or revolute joints; N_4 cylindrical or prismatic joints; N_5 rigid or deformable bodies, N_6 force actuators and N_7 virtual work elements.

The linear graph representation of a mechanical system is most easily described by means of an example. Consider the simple planar dynamic system, shown in figure 1(a), which consists of a body with center of gravity located at C.G. acted upon by two springs and a dashpot. Assuming negligible weight, the vector-network diagram, including the body traced in dashed lines to make the network easier to identify, is depicted in figure 1(b) and portrays a linear graph with six vertices and eight edges.

The edge e_1 is the inertial displacement vector representing the center of gravity, e_2 and e_3 define rigid or flexible arm elements, e_4 and e_5 specify displacement drivers while e_6 and e_7

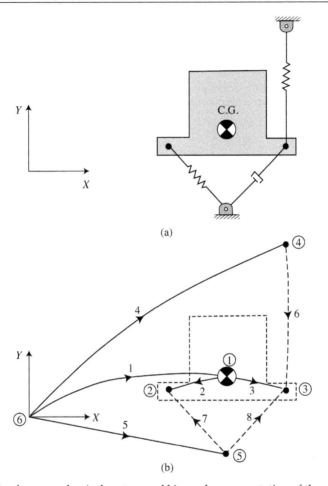

Fig. 1. a) Simple planar mechanical system and b) graph representation of the system

represent the springs and e_8 the dashpot. All vectors such as e_1, e_4 and e_5, the properties of which are established from an inertial frame, must emanate from the single ground node (*datum*) since that node is the only absolute fixed reference in the diagram, as compared to e_2 and e_3 which are defined relative to the body. Visibly, edges e_6, e_7 and e_8 span their respective two points of interconnection in the system.

It is possible to determine the number of branches and chords in a given graph $Graph(v, \epsilon)$ by applying basic theorems of graph theory (Andrews, 1977). Intuitively, a connected graph with v vertices will have $(v - 1)$ branches in its tree since a tree is a minimally connected graph. Consequently, by subtracting the number of branches from the total set of edges ϵ, there will be $(\epsilon - v + 1)$ chords in the cotree.

Given a graph with v vertices and ϵ edges, the order of interconnection of a system can be summarized in a v by ϵ incidence matrix. It is easy to construct since each edge is adjacent to exactly two vertices. The incidence matrix of $Graph(v, \epsilon)$ is denoted by $[\kappa]$ and is defined as follows: (1) each of the ν rows corresponds to a vertex of $Graph(\nu,\epsilon)$, (2) each of the ϵ columns corresponds to an edge of $Graph(\nu,\epsilon)$, and (3) entries $\kappa_{ij} = -1$ if the ith node is the initial vertex of the jth edge, $\kappa_{ij} = +1$ if the ith node is the final vertex of the jth edge, and $\kappa_{ij} = 0$ otherwise. All columns contain exactly two non-zero entries and $(\nu - 2)$ zero entries. The incidence matrix can always be compiled from inspection of the graph. As an illustration, the graph of figure 1(b) has the following incidence matrix:

$$[\kappa] = \begin{bmatrix} 1 & -1 & -1 & 0 & 0 & 0 & 0 & 0 \\ 0 & 1 & 0 & 0 & 0 & 0 & 1 & 0 \\ 0 & 0 & 1 & 0 & 0 & 1 & 0 & 1 \\ 0 & 0 & 0 & 1 & 0 & -1 & 0 & 0 \\ 0 & 0 & 0 & 0 & 1 & 0 & -1 & -1 \\ -1 & 0 & 0 & -1 & -1 & 0 & 0 & 0 \end{bmatrix} \tag{1}$$

From the incidence matrix, one can generate a cutset matrix. A cutset is a disconnecting set of edges such that after removal of that set of edges, the graph is divided into two or more components. A fundamental cutset (f-cutset) is considered a minimal cutset. A f-cutset reduces a connected graph into exactly two components. The selection of a tree branch will always identify one f-cutset associated to the branch and if a single cotree chord of the f-cutset is re-introduced into the graph, it unites the two residual subgraphs into a connected graph. As opposed to trees which represent a minimal set of edges which connect all the vertices of $Graph(\nu,\epsilon)$, a f-cutset is a minimal set of edges which disconnect some vertices from others.

Since a f-cutset exists for each tree branch, there will be $(\nu - 1)$ f-cutsets in a graph with ν vertices. Therefore, all f-cutsets can be assembled in a $(\nu - 1)$ by ϵ cutset matrix $[D]$ identifying all cotree terminal graphs acting through each vertex. For example, the cutset matrix of the diagram sketched in figure 1(b) can be obtained by simple row operations (Andrews, 1971) performed on the $(\nu - 1)$ rows of the incidence matrix. In this case, rows (2) and (3) are added to row (1) leading to the five branches cutset matrix,

$$[U_t \ D] = \begin{bmatrix} 1 & 0 & 0 & 0 & 0 & | & 1 & 1 & 1 \\ 0 & 1 & 0 & 0 & 0 & | & 0 & 1 & 0 \\ 0 & 0 & 1 & 0 & 0 & | & 1 & 0 & 1 \\ 0 & 0 & 0 & 1 & 0 & | & -1 & 0 & 0 \\ 0 & 0 & 0 & 0 & 1 & | & 0 & -1 & -1 \end{bmatrix} \tag{2}$$
$$\qquad\qquad\quad \text{tree} \qquad\qquad \text{cotree}$$

where $[U_t]$ is a $(\nu - 1)$ by $(\nu - 1)$ unit sub-matrix associated with tree branches and $[D]$ is a $(\nu - 1)$ by $(\epsilon - \nu + 1)$ sub-matrix associated with cotree chords. Since a f-cutset isolates a part of the system, its application to dynamic systems will become obvious when solving the relationship among forces which requires the construction of a "free-body diagram".

From the cutset matrix, one can generate a circuit matrix. A circuit is a connected subgraph in which exactly two edges are incident with each vertex. This concept will be molded into graph-networks through the use of a fundamental-circuit (f-circuit) which is defined as a subgraph which contains a single cotree chord, forming a closed chain, with tree branches. Following this definition, each chord of the cotree will form a f-circuit, thereby producing an independent set of closed chains since at least one edge will not be found in any other circuit. Earlier, the exact number of cotree chords in a $Graph(\nu,\epsilon)$ was found to be $(\epsilon - \nu + 1)$.

Combining this result with the fact that for each chord there is a corresponding f-circuit, the total number of independent circuits in a $Graph(v, \epsilon)$ is $(\epsilon - v + 1)$. The total set of f-circuits can be generated in a $(\epsilon - v + 1)$ by ϵ circuit matrix $[E]$ specifying each inertial terminal graph compatible with each circuit. The circuit matrix can be obtained by applying another basic theorem (Andrews, 1971) of graph theory which proves that the cutset and circuit matrices are **orthogonal**. This orthogonal relationship states that the scalar product of the cutset matrix and circuit matrix must vanish. Hence, in matrix notation, the submatrices $[D]$ and $[E]$ are related by the negative transpose principle of orthogonality (Andrews & Kesavan, 1975; Richard, 1985) (not well-known in dynamics):

$$[E] = -[D]^T \ or \ [D] = -[E]^T. \tag{3}$$

This liaison regulates the entire structure of this GT formulation. This principle can be demonstrated by using the sample graph of figure 1(b). Relation (3) can now be exploited to automatically transform the cutset matrix into the three chords circuit matrix,

$$[E \ U_c] = \begin{bmatrix} -1 & 0 & -1 & 1 & 0 & | & 1 & 0 & 0 \\ -1 & -1 & 0 & 0 & 1 & | & 0 & 1 & 0 \\ -1 & 0 & -1 & 0 & 1 & | & 0 & 0 & 1 \end{bmatrix} \tag{4}$$

$$\underbrace{}_{\text{tree}} \ \underbrace{}_{\text{cotree}}$$

where $[U_c]$ is a $(\epsilon - v + 1)$ by $(\epsilon - v + 1)$ unit sub-matrix associated with cotree chords and $[E]$ is a $(\epsilon - v + 1)$ by $(v - 1)$ sub-matrix associated with tree branches. In mechanical systems, the limitations to the freedom of movement of a body are specified by certain compatibility criteria which can be extricated from the circuit matrix. This matrix will prove to be a necessary mathematical tool in the resolution of dynamic systems since it provides some internal information relevant to the geometry of contact between terminal graphs.

At this point, one must introduce a physical meaning to these matrices. Unlike the traditional approach to dynamics in which Newton's second law is the basic postulate, in the graphical method there are two equally-important fundamental postulates: the vertex and circuit postulates. The vertex postulate states that the algebraic sum of through-variables $\{TV\}$, symbolizing quantities such as force $\mathbf{F}$, torque $\mathbf{T}$ and virtual work δW, corresponding to all the vectors incident with any vertex of the graph is, identically, zero. Essentially, this is recognizable as the dynamic force-balance law or d'Alembert's principle which requires that force summation upon each body, including d'Alembert's inertial variable, must be equal to zero. However, in this form it is more general, since it applies to any physical system; for example, in electrical systems, it is equivalent to Kirchhoff's current law. A **cutset** isolates a part of the system and is equivalent to the construction of a free-body-diagram. It has been shown earlier that these cutset equations are obtainable by simple row operations on the incidence matrix, and can be written in the form:

$$[U_t \ D]\{TV\} = 0 \tag{5}$$

where $[D_{ij}]$ is a sub-matrix containing +1, -1 and 0 depending whether element N_j is incident to N_i and sub-matrix U_t is a unit matrix associated tree elements. The column matrix $\{TV\}$ represents the through-variables ($\mathbf{F}_i$, $\mathbf{T}_i$ or δW_i) applied by element N_j. For the example depicted in figure 1(a), by combining equations (2) and (5), the vertex equation for node 1 representing the center of mass of the body is

$$F_1 + F_6 + F_7 + F_8 = 0. \tag{6}$$

In this case, conventional through variables, such as force vectors, are introduced in equation (6) (where $F_1 = -m\ddot{r}$ represents the inertial force).

The other governing postulate is the **circuit** postulate which states that the algebraic sum of across-variables $\{AV\}$, representing those quantities such as displacements r, velocities v, accelerations $\dot{v}$, virtual angular displacement $\delta\theta$, angular velocities ω and angular accelerations $\dot{\omega}$, corresponding to all the vectors included in any circuit is, identically, zero. Basically, this postulate represents the geometrical relations guiding the motion of mechanical systems. To be precise, each circuit equation alludes to a closed vector polygon respecting the geometric fit or compatibility law of rigid body dynamics. From graph theory, it can be shown that these kinematic constraint equations can be obtained from the cutset equations and are usually written in the form:

$$[E \ U_c]\{AV\} = 0 \tag{7}$$

where sub-matrix $[E_{ji}]$ specifies which element N_i is included in the closed loop N_j and sub-matrix U_c is a unit matrix associated with cotree elements. The column matrix $\{AV\}$ represents the across-variables (r_i, v_i, $\dot{v}_i$, $\delta\theta_i$, ω_i and $\dot{\omega}_i$) for element N_i. Since there is one independent circuit equation for each chord in the graph, the circuit equation for edge e_8 is

$$-r_1 - r_3 + r_5 + r_8 = 0 \tag{8}$$

where r represents the translational displacement of the corresponding element.

Consider the planar four-bar mechanism shown in figure 2(a). The system consists of three moving bodies (plus one fixed link) and four revolute joints. The link OA that is connected to the power source is called the input link (or crank). A driving torque causes the crank to rotate with angular velocity ω.The output link connects the moving pivot B to the ground pivot C. The coupler link connects the two moving pivots, A and B. The linear graph representation of the planar flexible four-bar linkage is depicted in figure 2(b) where the edges shown in bold comprise the spanning tree.

In this GT model, nodes (or vertices) are used to represent reference frames in the system while edges (lines) represent physical elements that connects these frames. Edges e_{51}, e_{52} and e_{53} represent the bodies in the system and e_{11}, e_{12}, both rigid arm elements, representing the body-fixed locations of the revolute joints at O and A, respectively. Edges e_{13} and e_{14} model the flexible links. The four revolute joints are modelled with joint edges e_{3i} ($i = 1, ..., 4$). Edge e_{21} represents a position driver and e_{61} is the driving torque. Finally, seven edges e_{7i} ($i = 1, ..., 7$) corresponding to dependent virtual work elements are automatically added to the system graph to complete the virtual topological equations.

Since we can also use virtual work δW as a through variable, the cutset equation for the tree joint element e_{34}, for example, is

$$\delta W_{33} + \delta W_{75} + \delta W_{76} + \delta W_{53} + \delta W_{34} = 0 \tag{9}$$

where each term corresponds directly to a physical component. Since edge e_{33} represents only relative displacements, it is necessary to introduce the dependent virtual work elements e_{74} and e_{75} to correctly capture the virtual works done by joint B on the coupler and output links. In general, whenever two bodies are connected by a component, two dependent virtual

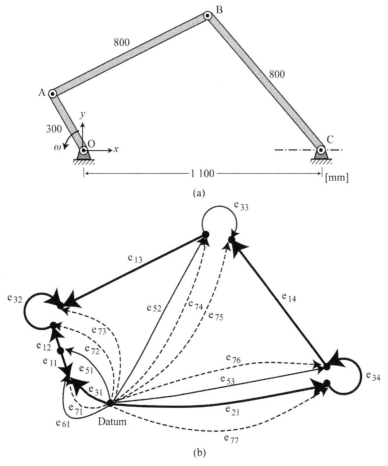

Fig. 2. a) Planar four-bar mechanism and b) graph representation of the system

work elements should automatically be included in the diagram model. Note that $\delta W_{33}=0$, since a frictionless joint cannot add or remove energy from a system. Furthermore, the sum of the virtual works done by joint B must also be zero, hence, $\delta W_{74} + \delta W_{75} = 0$. This fact can be exploited in this GT formulation to eliminate all joint reactions from the constitutive equations. Later, it will be shown that joint reactions can be retained in the GT algorithm by means of Lagrange multipliers.

The second set of topological equations, the circuit equations, can be generated automatically from the cutset equations. As an example, the circuit equation for cotree edge e_{33} is

$$r_{33} - r_{14} - r_{34} - r_{21} + r_{31} - r_{11} + r_{12} + r_{32} - r_{13} = 0 \qquad (10)$$

Clearly, this linear equation represents the vector loop closure condition for the circuit containing e_{33} where $r_{31} = r_{32} = r_{33} = r_{34} = 0$. Thus, the circuit equation (7) represent a set of linearly independent equations, one for each chord in the cotree.

3. Virtual work terminal equations for rigid bodies

The basic premise of the graphical approach is that each system element is modelled separately by defining its characteristics, independent of the other system elements, in the form of a "virtual terminal" (or constitutive) equation. Associated with every edge is one or more terminal equations that define the generalized force Q of an element corresponding to its branch coordinate q. These terminal equations are written in terms of the system through and across variables. In this formulation, a translational and rotational network is required in order to ensure consistency in the initial theory. Hence, these equations are functions of virtual displacements and virtual work.

In this formulation a rigid-arm element is used to represent the relative position and orientation of two reference frames on the same body. This second reference frame may be needed to locate a point of application for some other element. The terminal equations, in terms of virtual displacements, for a rigid-arm element are

$$\delta\boldsymbol{\theta}_1 = 0 \tag{11}$$

$$\delta\mathbf{r}_1 = \delta\boldsymbol{\theta}_5 \times \mathbf{r}_1 \tag{12}$$

where $\mathbf{r}_1$ is the rigid-arm position vector which is function of rotational body-fixed frame $\boldsymbol{\theta}_5$ since the direction of $\mathbf{r}_1$ varies as the rigid body rotates.

If the virtual work is the work done by specified forces on virtual displacements which are consistent with the constraints, with all other coordinates being kept constant, then the virtual work of force $\mathbf{F}$ is

$$\delta W = \delta\mathbf{r}^T\mathbf{F} \tag{13}$$

Now, let us consider a system whose kinematics are parameterized by a branch coordinate vector $\mathbf{q}$ where $\mathbf{r} = \mathbf{r}(\mathbf{q})$, then a virtual displacement $\delta\mathbf{r}$ is related to a branch coordinate variation $\delta\mathbf{q}$ by

$$\delta\mathbf{r} = \mathbf{r}_q\delta\mathbf{q} \tag{14}$$

where the subscript $\mathbf{q}$ indicates a partial derivative with respect to vector $\mathbf{q}$. Then the virtual work of a force may be obtained by substituting equation (14) in equation (13),

$$\delta W = \delta\mathbf{q}^T\mathbf{r}_q^T\mathbf{F} \tag{15}$$

Equation (13) may be interpreted as the scalar product of a branch variation $\delta\mathbf{q}$ and its generalized force $\mathbf{Q}$,

$$\delta W = \delta\mathbf{q}^T\mathbf{Q} \tag{16}$$

where the generalized force is defined as $\mathbf{Q} \equiv \mathbf{r}_q^T\mathbf{F}$. In this formulation, one may separate forces into applied forces $\mathbf{F}^A$ and constraint forces $\mathbf{F}^C$ such that,

$$\delta W = \delta\mathbf{q}^T(\mathbf{Q}^A + \mathbf{Q}^C) = \delta W^A + \delta W^C \tag{17}$$

If branch coordinates are consistent with constraints, then $\mathbf{q}$ is a vector with independent coordinates. Note that if constraints that act on the system are workless and if the branch coordinates $\mathbf{q}$ must satisfy constraint equations of the form

$$\boldsymbol{\Phi}(\mathbf{q}, t) = 0, \tag{18}$$

they are dependent. Thus branch coordinate variations must satisfy

$$\boldsymbol{\Phi_q} \delta\mathbf{q} = 0. \tag{19}$$

where $\boldsymbol{\Phi_q}$ is the Jacobian matrix of the constraint equation (18). If constraint forces are workless, it can be proven that

$$\delta W^C = \delta\mathbf{q}^T \mathbf{Q}^C = \delta\mathbf{q}^T \boldsymbol{\Phi_q^T} \lambda \tag{20}$$

where the Lagrange multiplier λ represents vector-generalized constraint forces.

Up to now, we have assumed that the reaction forces due to constraints neither produce nor consume work. This restriction has prohibited us from analyzing systems with non-rigid links (like elastic bodies which will be treated in the next section) or systems with energy-dissipating devices such as viscous or Coulomb dampers. The effect of such constraints is to produce certain pairs of internal forces which may be treated exactly as we would any other applied forces. Hence, the new form of terminal equations for applied forces has been written in the first part of equation (17). For elements containing physical constraints the terminal equation (20) is introduced. Care must be taken when applying the constraint generalized force because the physical constraints do no work and its Lagrange multiplier form must be introduced in the initial cutset equations at the beginning of the substitution procedure.

Spring and gravitational forces belong to the wider class of so-called conservative forces for which the generalized force can be efficiently calculated from the potential energy. Thus, if forces that act on a system can be expressed as the gradient of a scalar function of generalized coordinates, the virtual work may be written as the negative of a variation in potential energy V and the new terminal equation for conservative forces becomes

$$\delta W^{cons} = \delta\mathbf{q}^T \mathbf{Q}^{cons} = -\delta\mathbf{q}^T V_\mathbf{q}^T \tag{21}$$

Now, let us consider a moving system defined by a set of branch coordinates. If between these branch coordinates and time t there exists some relation of the form of equation (18), it is said that the system is moving under constraint. This means that the functions of constraints are geometric or kinematic conditions which restrain the possibilities of motion of the system. For a spatial mechanical system there is a limited number of types of functions of constraint, represented by the joints between the bodies. Figure 3 presents the four most common types of ideal joints which can be found in multibody systems.

A spherical joint shown in figure 3(a) is defined by the condition that the center of the ball coincides with the center of the socket. The terminal equations are, then, 3 scalar constraint equations that restricts the relative positions between the bodies,

$$[\mathbf{e}^x \ \mathbf{e}^y \ \mathbf{e}^z]^T \delta\mathbf{r}_3 = [0 \ 0 \ 0]^T \tag{22}$$

where $\mathbf{e}^x, \mathbf{e}^y, \mathbf{e}^z$ are unit vectors and $\delta\mathbf{r}_3$ represents the relative virtual displacement between the bodies. Essentially, the three constraining equations (22) impose that there is no relative displacements at the joint in the $x - y - z$ directions.

A revolute joint, shown in figure 3(b), is constructed with bearings that allow relative rotation about a common axis in a pair of bodies, but precludes relative translation along this axis. The

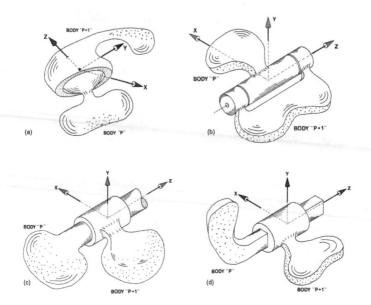

Fig. 3. Ideal kinematic joints a) spherical, b) revolute, c) cylindrical and d) prismatic

five terminal constrained equations are

$$[\mathbf{e}^x \ \mathbf{e}^y \ \mathbf{e}^z]^T \delta \mathbf{r}_3 = [0 \ 0 \ 0]^T \quad ; \quad [\mathbf{e}^x \ \mathbf{e}^y]^T \delta \boldsymbol{\theta}_3 = [0 \ 0]^T \tag{23}$$

where $\delta \boldsymbol{\theta}_3$ represents the relative virtual angular rotation between the bodies.

A cylindrical joint between a pair of bodies is shown in figure 3(c). It permits relative translation and relative rotation between bodies about a common axis. The four terminal constrained equations are

$$[\mathbf{e}^x \ \mathbf{e}^y]^T \delta \mathbf{r}_4 = [0 \ 0]^T \quad ; \quad [\mathbf{e}^x \ \mathbf{e}^y]^T \delta \boldsymbol{\theta}_4 = [0 \ 0]^T \tag{24}$$

A prismatic joint, shown in figure 3(d), allows relative translation along a common axis between a pair of bodies, but precludes relative rotation about this axis. The five terminal constrained equations are

$$[\mathbf{e}^x \ \mathbf{e}^y]^T \delta \mathbf{r}_4 = [0 \ 0]^T \quad ; \quad [\mathbf{e}^x \ \mathbf{e}^y \ \mathbf{e}^z]^T \delta \boldsymbol{\theta}_4 = [0 \ 0 \ 0]^T \tag{25}$$

At this point, one can represent these kinematic joints in a (6×6) Boolean matrix,

$$\mathbf{E}^k = diag(\mathbf{e}^k) = \begin{bmatrix} \mathbf{E}_t^k & 0 \\ 0 & \mathbf{E}_r^k \end{bmatrix} \tag{26}$$

where $\mathbf{E}_t^k$ and $\mathbf{E}_r^k$ are two diagonal (3×3) tensors which define the translational and rotational connexions between bodies.

Let us now consider rigid body elements N_5. The graph for this type of element consists of an edge e_5 from the datum node to a local reference frame on the body. The use of the variational form of Lagrange's equation relies upon a correct formulation of the kinetic energy of the multibody system in terms of branch coordinates. If the system is composed of n_b rigid bodies, the kinetic energy of the ith body is defined as

$$T_i = \frac{1}{2} m_i \mathbf{v}_i^T \mathbf{v}_i + \frac{1}{2} \boldsymbol{\omega}_i^{\prime T} [I_i'] \boldsymbol{\omega}_i' \tag{27}$$

where m_i is the mass of body i, $\mathbf{v}_i$ is the velocity vector of the centre of mass of the body in inertial space, $[I_i']$ is the inertia tensor of the body expressed in a body-fixed ' coordinate system and $\boldsymbol{\omega}_i'$ is the angular velocity vector of the body in inertial space and expressed in the inertia ' tensor coordinate system. Note that absolute or inertial-space velocities must be used although they may be expressed in any convenient (inertial or non-inertial) coordinate system. Then, the virtual work terminal equation for the ith body becomes

$$\delta W_{5i} = \delta \mathbf{q}^T \left\{ \left[\frac{d}{dt} \left(\frac{\partial T_i}{\partial \dot{\mathbf{q}}} \right) - \frac{\partial T_i}{\partial \mathbf{q}} \right]^T \right\} \tag{28}$$

It is assumed at this point that the velocities $\{\mathbf{v}_i\}$ and $\{\boldsymbol{\omega}_i\}$ have been found in symbolic form for each body component in terms of branch coordinates and their derivatives. In order to apply the variational form of Lagrange's equation, the terms $\frac{d}{dt}\left(\frac{\partial T_i}{\partial \dot{\mathbf{q}}}\right)$ and $\frac{\partial T_i}{\partial \mathbf{q}}$ will be required for each branch coordinates. By partial differentiation of equation (27),

$$\frac{\partial T_i}{\partial \mathbf{q}} = \left(m_i \mathbf{v}_i^T \mathbf{v}_{i\mathbf{q}} + \boldsymbol{\omega}_i^{\prime T} [I_i'] \boldsymbol{\omega}_{i\mathbf{q}}' \right) \tag{29}$$

and similarly,

$$\frac{\partial T_i}{\partial \dot{\mathbf{q}}} = \left(m_i \mathbf{v}_i^T \mathbf{v}_{i\dot{\mathbf{q}}} + \boldsymbol{\omega}_i^{\prime T} [I_i'] \boldsymbol{\omega}_{i\dot{\mathbf{q}}}' \right) \tag{30}$$

The time derivative of equation (30) is

$$\frac{d}{dt}\left(\frac{\partial T_i}{\partial \dot{\mathbf{q}}}\right) = \left(m_i \dot{\mathbf{v}}_i^T \mathbf{v}_{i\dot{\mathbf{q}}} + m_i \mathbf{v}_i^T \dot{\mathbf{v}}_{i\dot{\mathbf{q}}} + \dot{\boldsymbol{\omega}}_i^{\prime T} [I_i'] \boldsymbol{\omega}_{i\dot{\mathbf{q}}}' + \boldsymbol{\omega}_i^{\prime T} [I_i'] \dot{\boldsymbol{\omega}}_{i\dot{\mathbf{q}}}' \right) \tag{31}$$

Substituting equations (29) and (31) into the variational form of Lagrange's equation (28) gives

$$\delta W_{5i} = \delta \mathbf{q}^T \left\{ \left[m_i \dot{\mathbf{v}}_i^T \mathbf{v}_{i\dot{\mathbf{q}}} + \dot{\boldsymbol{\omega}}_i^{\prime T} [I_i'] \boldsymbol{\omega}_{i\dot{\mathbf{q}}}' + m_i \mathbf{v}_i^T \mathbf{P}_{i/\mathbf{q}}^v + \boldsymbol{\omega}_i^{\prime T} [I_i'] \mathbf{P}_{i/\mathbf{q}}^{\omega'} \right]^T \right\} \tag{32}$$

where

$$\mathbf{P}_{i/\mathbf{q}}^v = \dot{\mathbf{v}}_{i\dot{\mathbf{q}}} - \mathbf{v}_{i\mathbf{q}} \tag{33}$$

$$\mathbf{P}_{i/\mathbf{q}}^{\omega'} = \dot{\boldsymbol{\omega}}_{i\dot{\mathbf{q}}}' - \boldsymbol{\omega}_{i\mathbf{q}}' \tag{34}$$

The quantity $\mathbf{P}_{i/\mathbf{q}}^v$ can be shown to be equal to zero. Since $\mathbf{r}_i = \mathbf{r}_i(\mathbf{q}, t)$, we have

$$\mathbf{v}_i = \dot{\mathbf{r}}_i = \mathbf{r}_{i\mathbf{q}} \dot{\mathbf{q}} + \mathbf{r}_{it} \tag{35}$$

From equation (35), form the partial derivative of $\mathbf{v}_i$ with respect to $\dot{\mathbf{q}}$

$$\mathbf{v}_{i\dot{\mathbf{q}}} = \mathbf{r}_{i\mathbf{q}} \tag{36}$$

Take the time derivatives of both sides of equation (36) to get

$$\dot{v}_{i\dot{q}} = \dot{r}_{iq} = v_{iq} \tag{37}$$

Substitute equation (37) into equation (33) to show that $\mathbf{P}^v_{i/q} = 0$. When the angular velocity vector is the exact derivative of another vector function of branch coordinates and time, the preceding argument will show that $\mathbf{P}^{\omega'}_{i/q} = 0$. This is always the case for problems formulated in one or two-dimensional space where angular velocities are simply the time derivatives of angular coordinates. In problems which must be formulated in three dimensions, finite rotations cannot be represented as vectors and $\mathbf{P}^{\omega'}_{i/q}$ is generally non-zero.

The virtual equation (32) is now simplified to

$$\delta W_{5i} = \delta\mathbf{q}^T \left\{ \left[m_i \dot{\mathbf{v}}_i^T \mathbf{v}_{i\dot{q}} + \dot{\omega}_i'^T [I_i']\omega_{i\dot{q}}' + \omega_i'^T [I_i']\mathbf{P}^{\omega'}_{i/q} \right]^T \right\} \tag{38}$$

The right hand side of equation (38) can be separated into terms containing branch accelerations $\ddot{\mathbf{q}}$ and terms containing products of branch velocities $\dot{\mathbf{q}}\dot{\mathbf{q}}$. This is necessary in order to place the coefficients of the terms $\ddot{\mathbf{q}}$ into an augmented mass matrix $\hat{\mathbf{M}}_i$ and the product terms into a generalized force $\mathbf{Q}_i^K$. Write the time-derivatives of the velocity vectors as

$$\dot{\mathbf{v}}_i = \mathbf{v}_{i\dot{q}}\ddot{\mathbf{q}} + \dot{\mathbf{v}}_i^p \tag{39}$$

$$\dot{\omega}_i = \omega_{i\dot{q}}'\ddot{\mathbf{q}} + \dot{\omega}_i'^p \tag{40}$$

The terms superscripted ^{p} contain all products of generalized velocities. Then, exploit equations (39) and (40) to rewrite equation (38) into the final terminal equation for rigid bodies as follows

$$\delta W_{5i} = \delta\mathbf{q}^T \left\{ \left[\hat{\mathbf{M}}_i \ddot{\mathbf{q}} + \mathbf{Q}_i^K \right]^T \right\} \tag{41}$$

where

$$\hat{\mathbf{M}}_i = \left(m_i \mathbf{v}_{i\dot{q}}^T \mathbf{v}_{i\dot{q}} + \omega_{i\dot{q}}'^T [I_i']\omega_{i\dot{q}}' \right) \tag{42}$$

and

$$\mathbf{Q}_i^K = \left(m_i \dot{\mathbf{v}}_i^{pT} \mathbf{v}_{i\dot{q}} + \dot{\omega}_i'^{pT} [I_i']\omega_{i\dot{q}}' + \omega_i'^T [I_i']\mathbf{P}^{\omega'}_{i/q} \right) \tag{43}$$

The elements of the coefficient matrix $\hat{\mathbf{M}}_i$ are the coefficient of $\ddot{\mathbf{q}}$ appearing in the differential equation corresponding to the branch coordinate $\mathbf{q}$. The forcing vector $\mathbf{Q}_i^K$ shown in equation (43) is made up of all the terms in the equations of motion which do not contain second derivatives. This vector contains part of the contribution of the kinetic energy to the equations of motion. Substituting the general form of virtual work terminal equations for each element in the cutset equation for this tree body constraint or joint, one gets

$$\delta\mathbf{q}^T \left\{ \hat{\mathbf{M}} \ddot{\mathbf{q}} + \mathbf{Q}^K + \mathbf{Q}^C + \mathbf{Q}^{cons} - \mathbf{Q}^A \right\} = 0 \tag{44}$$

This variational equation of motion holds for arbitrary virtual displacement $\delta\mathbf{q}$, so it is equivalent to the Lagrange multiplier form of constrained equations of motion.

Together, the cutset, circuit and terminal equations form a necessary and sufficient set of motion equations for determining the time response of multibody systems. However, an efficient approach to this problem consists in reducing the number of equations that need to be solved simultaneously by using branch transformation equations for a tree selection. The first step consists in defining the problem by creating a proper spanning tree and generating the branch transformation equations for all the elements in the cotree. These transformations will be used to replace the across-variables for cotree elements as function of branch coordinates.

Depending on the topology of the mechanical system and the specified tree, the branch coordinates may not be independent quantities. If the number of coordinates n is greater than the degrees of freedom f, then $(c = n - f)$ constraint equations are required to express the dependency between coordinates. The constraint equations are obtained directly from the joints and motion drivers in the cotree, by projecting their circuit equations onto the joint reaction space (McPhee, 1998). Upon substitution of the branch transformation equations into these circuit equations, the constraint equations are obtained in the form of equation (18) which constitutes a set of c nonlinear algebraic equations. Differentiating twice equation (18), using the chain rule of differentiation, yields the c constraint acceleration equations

$$\mathbf{\Phi_q}\ddot{\mathbf{q}} = -\left[(\mathbf{\Phi_q}\dot{\mathbf{q}})_\mathbf{q}\dot{\mathbf{q}} + 2\mathbf{\Phi}_{\mathbf{q}t}\dot{\mathbf{q}} + \mathbf{\Phi}_{tt}\right] \equiv \mathbf{\Lambda} \qquad (45)$$

In general, the n branch coordinates are not independent, but are related by the c kinematic constraint equations (18). Thus, these constraint acceleration equations must be appended to the set of dynamic equations (44), giving $(n + c)$ equations to solve for branch coordinates $\mathbf{q}$ and the c reaction loads λ. Substituting equation (18) into equation (44) and combining with equation (45) finally yields the classical system differential-algebraic equations of motion for rigid multibody systems in matrix form,

$$\begin{bmatrix} \hat{\mathbf{M}} & \mathbf{\Phi}_\mathbf{q}^T \\ \mathbf{\Phi}_\mathbf{q} & 0 \end{bmatrix}\begin{Bmatrix} \ddot{\mathbf{q}} \\ \lambda \end{Bmatrix} = \begin{Bmatrix} \mathbf{Q}^{total} \\ \mathbf{\Lambda} \end{Bmatrix} \qquad (46)$$

where $\hat{\mathbf{M}}$ is a symmetric augmented $(n \times n)$ mass matrix, $\mathbf{\Phi}_\mathbf{q}$ is a $(c \times n)$ constraint Jacobian matrix and the total generalized force $\mathbf{Q}^{total} = \mathbf{Q}^A - \mathbf{Q}^K - \mathbf{Q}^{cons}$. Using equation (42) it is easy to form symbolically the coefficient matrix $\hat{\mathbf{M}}$, element by element, from partial derivatives of the velocity vectors. Non-linearities have been preserved throughout the formulation as each element may contain branch coordinates and quantities which vary with time. Note that the global matrix coefficient is symmetric. This requires only that the inertia tensor be symmetric, which it is. The off-diagonal mass sub-matrices represent coupling effects between adjacent bodies. The fundamental form of these equations and physical properties of kinetic energy guarantee that a unique solution of the constrained equations of motion exists.

4. Flexible multibody systems (FMS)

This variational graph-theoretical approach is based on the flexible models developped by (Shabana, 1986) and (Tennich, 1994). In this formulation a flexible arm element, as shown in figure 4, is used to represent the relative position and orientation of two reference frames on the same body. This second reference frame may be needed to locate a point of application for some other element.

A flexible arm element is defined as being made up of a continuum of particles that can move relative to one another. Since actual bodies are never perfectly rigid, small deformation effects are often added and can have great influence on the motion of mechanisms that are made up of multiple bodies. A flexible arm element can be located by defining the position vector $\mathbf{R}$ of the origin of its body-fixed frame $\Re'(X', Y', Z')$, relative to the global reference frame $\Re(XYZ)$.

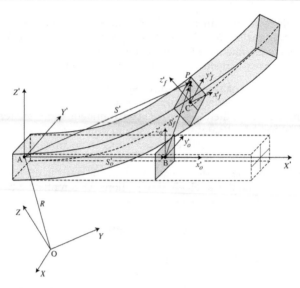

Fig. 4. Flexible arm element

The global location of an arbitrary point P on a flexible body can be described as

$$\mathbf{r} = \mathbf{R} + \mathbf{\Pi}\,\mathbf{s}' = \mathbf{R} + \mathbf{\Pi}\,(\mathbf{s}'_o + \mathbf{s}'_f) \tag{47}$$

where $\mathbf{\Pi}$ represents an Euler parameter transformation matrix (Wittenburg, 1977) from the structure $\mathfrak{R}'(X', Y', Z')$ coordinate system to the global reference frame $\mathfrak{R}(XYZ)$. The vector $\mathbf{s}'_o$ is the initial undeformed position of point p with respect to the body-fixed reference frame, $\mathbf{s}'_f$ represents the flexible displacement of point p and $\mathbf{s}'(= \mathbf{s}'_o + \mathbf{s}'_f)$ represents a flexible arm element N_1.

Since the body-fixed frame $\mathfrak{R}'(x', y', z')$ translates and rotates relative to the global frame, the vector $\mathbf{R}$ and Euler transformation matrix $\mathbf{\Pi}$ are functions of time. Both sides of eq.(47) may be differentiated twice with respect to time to obtain the acceleration equation,

$$\ddot{\mathbf{r}} = \ddot{\mathbf{R}} + \mathbf{\Pi}\,(\ddot{\mathbf{s}}' + \tilde{\dot{\omega}}\,\mathbf{s}' + \tilde{\omega}\,\tilde{\omega}\,\mathbf{s}' + 2\,\tilde{\omega}\,\dot{\mathbf{s}}') \tag{48}$$

where $\dot{\omega}$ and $\ddot{\mathbf{s}}'$ are, respectively, the angular acceleration of the body and the second derivative of the elastic displacement of point p and $\tilde{\omega}$ represents a 3×3 skew-symmetric matrix which performs a cross-product multiplication and represents the angular velocity vector of the body-fixed frame $\mathfrak{R}'$.

To describe the deformation of a flexible body, one can discretize the structure into finite elements. If a reference frame $\mathfrak{R}(e_1, e_2, e_3)$ is attached to the jth element, the displacement of point p on a flexible body can be given by

$$\mathbf{s}' = \mathbf{N}^j\,(\mathbf{s}'_{on} + \mathbf{s}'_{fn}) = \mathbf{N}^j\,\mathbf{s}''_n \tag{49}$$

with

$$\mathbf{N}^j = \mathbf{Q}^j\,\mathbf{N}^{je}\,\mathbf{T}^{j^T}$$

where $\mathbf{s}'_{on}$ and $\mathbf{s}'_{fn}$ are the non-deformed and the nodal visco-elastic displacement of point p and $\mathbf{s}_n^{'j}$ is the total displacement nodal vector of the jth element. Note that $\mathbf{N}^j$ is a finite element spatial interpolation function from the jth element to the body frame $\mathfrak{R}'$ where $\mathbf{Q}^j$ represents a transformation matrix from the frame $\mathfrak{R}(e_1, e_2, e_3)$ to the body reference frame $\mathfrak{R}(\bar{x}, \bar{y}, \bar{z})$ and $\mathbf{T}^j$ is a transfert matrix assembled with the $\mathbf{Q}^j$ matrices.

Figures 5(a) and 5(b) represent the 3-D beam and the triangular shell elements, respectively. For the 3-D beam element, the two spatial interpolation function is given

$$N_1^{je} = \frac{1}{2}(1 - \xi) \quad and \quad N_2^{je} = \frac{1}{2}(1 + \xi) \tag{50}$$

where ξ is an adimensional variable mesured along each beam element, with $\xi = -1$ at node 1 and $\xi = 1$ at node 2. For the triangular shell element, the three spatial interpolation functions are given by

$$N_1^{je} = 1 - \xi - \eta \quad ; \quad N_2^{je} = \xi \quad ; \quad N_3^{je} = \eta \tag{51}$$

where ξ and η are adimensional variables mesured along the triangular element.

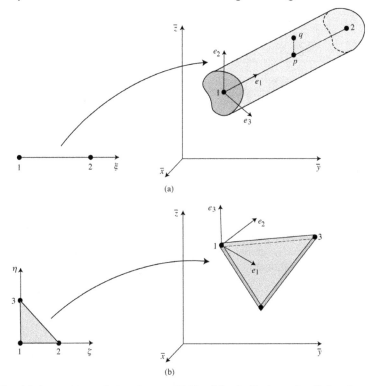

Fig. 5. a) Flexible beam linear finite element b) Flexible shell triangular finite element

Vectors $\mathbf{e}_1^j$, $\mathbf{e}_2^j$ and $\mathbf{e}_3^j$, in figure 5, form a reference surface base for the jth element which is independent of the nodal numbering. Hence, the transfert matrix between the frame $\mathfrak{R}(e_1, e_2, e_3)$ to the body reference frame $\mathfrak{R}(\bar{x}, \bar{y}, \bar{z})$ can be written as $\mathbf{Q}^j = (e_1, e_2, e_3)^j$. Then, the diagonal transfert matrix $\mathbf{T}^j$ between the element local frame $\Re(e_1, e_2, e_3)^j$ and the deformable body frame $\Re(\bar{x}, \bar{y}, \bar{z})$ can be written with six rotation tensors $\mathbf{Q}^j$ (of dimension 3×3).

Finally, the acceleration expression of point p on the flexible body can then be rewritten from the discretized form of $\mathbf{s}'$, from eq.(48),

$$\ddot{\mathbf{r}} = \ddot{\mathbf{R}} - \mathbf{\Pi} \, \bar{\mathbf{s}}' \, \mathbf{\Lambda} \, \ddot{\theta} + \mathbf{\Pi} \, N^j \, \ddot{\mathbf{s}}_n^{\prime j} + \mathbf{\Pi} \, (\bar{\omega} \, \bar{\omega} \, N^j \, \mathbf{s}_n^{\prime j} + 2 \, \bar{\omega} \, N^j \, \dot{\mathbf{s}}_n^{\prime j}) \tag{52}$$

where $\omega = \mathbf{\Lambda} \, \dot{\theta}$ with $\mathbf{\Lambda}$ representing an Euler transformation matrix for the angular velocity of the body with respect to the body reference frame $\Re'$. The generalized coordinate vector $\mathbf{q}$ for a flexible body can then be assembled from the position vector $\mathbf{R}$, orientation θ of the origin of the body reference frame $\Re'$ and the nodal coordinate vector $\mathbf{s}_n'$ with $\mathbf{q} = \{\mathbf{R}, \, \theta, \, \mathbf{s}_n'\}^T$.

Consider a volume element $dv = dx \, dy \, dz$ near point p on a deformable body. The constitutive virtual work equation of this volume element dv can be separated in three different virtual work components (Shi et al., 2001; Tennich, 1994),

$$\delta W_5 = \delta W_5^{inertial} - \delta W_5^{forces} + \delta W_5^{internal} = 0 \tag{53}$$

with

$$\delta W_5^{inertial} = \int_V \delta \mathbf{r}^T \rho \, \ddot{\mathbf{r}} \, dv \tag{54}$$

$$\delta W_5^{forces} = \int_V \delta \mathbf{r}^T \, \mathbf{f}_v \, dv \tag{55}$$

$$\delta W_5^{internal} = \int_V \delta \epsilon^T \, \sigma \, dv \tag{56}$$

where $\delta \epsilon$ is the column matrix of varied strain components in the local frame; $(\rho \, \ddot{\mathbf{r}})$ is the inertial force per unit volume; σ contains the corresponding internal stress components and $\mathbf{f}_v$ is the body volume forces including gravity.

For a deformable body, the virtual work of all inertial forces can be written from equation (54) under the following general form,

$$\delta W_5^{inertial} = \delta \mathbf{q}^T \, \mathbf{M} \, \ddot{\mathbf{r}} - \delta \mathbf{q}^T \, \mathbf{Q}^q \tag{57}$$

where $\mathbf{M}$ and $\mathbf{Q}^q$ represent the global mass matrix and the quadratic velocity vector including Coriolis term, respectively. The global mass matrix of the jth finite element of the flexible body can be assembled from the elementary mass matrices,

$$\mathbf{M}^j = \int_{V^j} \rho^j \begin{bmatrix} I & -\mathbf{\Pi} \, \bar{\mathbf{s}}' \, \mathbf{\Lambda} & \mathbf{\Pi} \, N^j \\ & \mathbf{\Lambda}^T \, \bar{\mathbf{s}}'^T \, \bar{\mathbf{s}}' \, \mathbf{\Lambda} & -\mathbf{\Lambda}^T \, \bar{\mathbf{s}}'^T \, N^j \\ Sym. & & N^{j^T} \, N^j \end{bmatrix} dv = \begin{bmatrix} \mathbf{M}_{RR} & \mathbf{M}_{R\theta} & \mathbf{M}_{Rf} \\ & \mathbf{M}_{\theta\theta} & \mathbf{M}_{\theta f} \\ Sym. & & \mathbf{M}_{ff} \end{bmatrix}^j \tag{58}$$

where ρ^j and V^j are, respectively, the mass density and the volume of the jth element. The elements of the mass matrix $\mathbf{M}$ are the coefficient of $\ddot{\mathbf{q}}$ appearing in the differential equation corresponding to the branch coordinate $\mathbf{q}$. Using equation (53), it is easy to assemble the coefficient matrix $\mathbf{M}$, element by element, from the transformation matrices between reference frames. The formulation of the time-invariant matrices $\mathbf{M}_{RR}$ and $\mathbf{M}_{ff}$

is straight-forward. The matrix $\mathbf{M}_{RR}$ is a diagonal matrix whose elements are equal to the mass of the element. The matrix $\mathbf{M}_{ff}$ is the conventional mass matrix that arises in any finite element analysis. Non-linearities have been preserved throughout the formulation as each element may contain branch coordinates and quantities which vary with time. The off-diagonal mass sub-matrices represent coupling effects between translational, rotational and flexible elements. The matrices $\mathbf{M}_{Rf}$ and $\mathbf{M}_{\theta f}$ represent the inertia coupling between gross body motion and small body deformation. These matrices, in addition to the inertia tensor $\mathbf{M}_{\theta\theta}$, are implicitly time dependent since they are functions of the body generalized coordinates.

In a similar fashion, the global quadratic velocity vectors can be assembled for the jth finite element of the flexible body,

$$
\left\{
\begin{array}{c}
Q_R^q \\
Q_\theta^q \\
Q_f^q
\end{array}
\right\}^j
= -\int_{V^j} \rho^j
\left\{
\begin{array}{c}
\Pi \\
\Lambda^T \bar{\mathbf{s}}'^{T} \\
N^{jT}
\end{array}
\right\}
[\tilde{\omega}\,\tilde{\omega}\,N\,\mathbf{s}'_n + 2\,\tilde{\omega}\,N\,\dot{\mathbf{s}}'_n]^j\,dv
\tag{59}
$$

The inertial forcing vector Q^q is made up of all the terms in the equations of motion which do not contain second derivatives.

The virtual work of body forces $\mathbf{f}_v$ for the jth finite element can also be written at once,

$$
\left(\delta W_6^{forces}\right)^j = \int_{V^j} \delta \mathbf{r}^T\,\mathbf{f}_v^j\,dv = [\delta \mathbf{R}^T \quad \delta\theta^T \quad \delta \mathbf{s}_n'^{jT}]
\left\{
\begin{array}{c}
\mathbf{F}_R \\
\mathbf{F}_\theta \\
\mathbf{F}_f
\end{array}
\right\}^j
\tag{60}
$$

Finally, the virtual work for internal constraints of the jth finite element was defined earlier,

$$
\left(\delta W_5^{internal}\right)^j = \int_{V^j} \delta \mathbf{e}^{jT}\,\sigma^j\,dv
\tag{61}
$$

where e^{jT} and σ^j represent, respectively, Cauchy's deformation and strain vectors. For a viscous elastic material governed by the Kelvin-Voigt model (Christensen, 1975; Flugge, 1967), the behaviour law between stress and strain is established as,

$$
\sigma^j = \mathbf{H}^j\,e^j + \mathbf{G}^j\,\dot{e}^j = \mathbf{H}^j\,\Xi^j\,\mathbf{s}_n'^j + \mathbf{G}^j\,\Xi^j\,\dot{\mathbf{s}}_n'^j
\tag{62}
$$

where $\mathbf{H}^j$ and $\mathbf{G}^j$ are, respectively, the elastic and viscous tensors for this behaviour law and are function of Young's modulus and poisson's coefficient. The matrix Ξ^j represents spatial interpolation deformation functions. Hence, the internal virtual work for the jth finite element can be written under the following form,

$$
\left(\delta W_5^{internal}\right)^j = \delta\,\mathbf{s}_n'^{jT}\,\mathbf{K}_{ff}^j\,\mathbf{s}_n'^j + \delta\,\mathbf{s}_n'^{jT}\,\mathbf{C}_{ff}^j\,\dot{\mathbf{s}}_n'^j
\tag{63}
$$

with

$$\mathbf{K}_{ff}^{j} = \int_{V^{j}} [\mathbf{\Xi}^{T} \, \mathbf{H} \, \mathbf{\Xi}]^{j} \, dv \tag{64}$$

$$\mathbf{C}_{ff}^{j} = \int_{V^{j}} [\mathbf{\Xi}^{T} \, \mathbf{G} \, \mathbf{\Xi}]^{j} \, dv \tag{65}$$

Equations (64) and (65) are, respectively, the stiffness matrix and the viscous damping matrix of the jth element. With the body kinetic and strain energy in hand, the virtual work can be exploited to generate the system equations of motion of a single flexible body. Hence, the general form of virtual work terminal equation for each body δW_{5i}, required in the cutset equation, can be written

$$\delta \mathbf{q}^{T} \{\mathbf{M} \, \ddot{\mathbf{q}} + \mathbf{C} \, \dot{\mathbf{q}} + \mathbf{K} \, \mathbf{q} - \mathbf{F} - \mathbf{Q}\} = 0 \tag{66}$$

This variational equation holds for arbitrary virtual displacement $\delta \mathbf{q}$, so the terms in parentheses are the well-known equations of motion in standard form. Hence, for each flexible body in the system, the translational, rotational and viscous-elastic equations of motion can now be assembled into a partitioned matrix formulation,

$$\begin{bmatrix} \mathbf{M}_{RR} & \mathbf{M}_{R\theta} & \mathbf{M}_{Rf} \\ & \mathbf{M}_{\theta\theta} & \mathbf{M}_{\theta f} \\ Sym. & & \mathbf{M}_{ff} \end{bmatrix} \begin{Bmatrix} \ddot{\mathbf{R}} \\ \ddot{\boldsymbol{\theta}} \\ \ddot{\mathbf{s}}_{n}' \end{Bmatrix} + \begin{bmatrix} 0 & 0 & 0 \\ 0 & 0 & 0 \\ 0 & 0 & \mathbf{C}_{ff} \end{bmatrix} \begin{Bmatrix} \dot{\mathbf{R}} \\ \dot{\boldsymbol{\theta}} \\ \dot{\mathbf{s}}_{n}' \end{Bmatrix} + \begin{bmatrix} 0 & 0 & 0 \\ 0 & 0 & 0 \\ 0 & 0 & \mathbf{K}_{ff} \end{bmatrix} \begin{Bmatrix} \mathbf{R} \\ \boldsymbol{\theta} \\ \mathbf{s}_{n}' \end{Bmatrix} = \begin{Bmatrix} \mathbf{FQ}_{R}^{q} \\ \mathbf{FQ}_{\theta}^{q} \\ \mathbf{FQ}_{f}^{q} \end{Bmatrix} \tag{67}$$

where $\mathbf{FQ}_{i}^{q} = \mathbf{F}_{i} + \mathbf{Q}_{i}^{q}$ for $(i = R, \theta, f)$. Since the total virtual work of the system of flexible bodies is the sum of the individual virtual work, the algorithm must incorporate the sum of all bodies in the formulation using a global cutset equation. Then, the variational equations of motion for flexible multibody systems may be obtained from the tree joint element which connects the whole system of bodies to the ground body through a path consisting entirely of branches. To get the contribution of all physical components to the system, the terminal virtual work equations are written for all bodies in the tree. By summing the cutset equation, similar to eq.(5), for all tree flexible bodies, the contribution of all physical components to the system virtual work equations are captured. The dynamic form of these expressions and fundamental properties of virtual work guarantee that a unique solution of the flexible set of equations of motion exists.

Depending on the topology of the mechanical system and the specified tree, the branch coordinates may not be independent quantities. If the number of coordinates is greater than the degrees of freedom, then constraint equations are required to express the dependency between coordinates. The constraint equations are obtained directly from the joints and motion drivers in the cotree, by projecting their circuit equations, similar to eq.(7), onto the joint reaction space. Then, if kinematic joint k establishes a translational connexion between points a and b located on bodies i and j, the following vectorial constraint expression must be respected,

$$\mathbf{\Pi}^{k \, T} [\mathbf{r}^{i} - \mathbf{r}^{j}] + \mathbf{r}_{k}'^{k} = 0 \tag{68}$$

where $\mathbf{r}^{i}$ and $\mathbf{r}^{j}$ are the position vectors of the fixation points of the kinematic joint k on bodies i and j while $\mathbf{r}_{k}'^{k}$ represents the length of the joint. In a similar fashion, the rotational connexion at kinematic joint k between points a and b can be written,

$$(\boldsymbol{\omega}^i + \dot{\boldsymbol{\alpha}}^i)_k + \boldsymbol{\omega}_k^k - (\boldsymbol{\omega}^j + \dot{\boldsymbol{\alpha}}^j)_k = 0 \tag{69}$$

where $\boldsymbol{\omega}^i$, $\boldsymbol{\omega}^j$ and $\boldsymbol{\omega}_k^k$ are angular velocities of bodies i and j and kinematic joint k, expressed in the joint reference frame. By introducing the tensor $\mathbf{E}^k$, obtained in equation (26), to eliminate all superflous constraints and substituting the relations for the derivatives of the transformation matrices, the translational and rotational geometrical constraint equations (68) and (69) can be derived twice with respect to time to obtain,

$$\mathbf{E}^k \mathbf{\Pi}^{k\,T} \begin{bmatrix} I & -\mathbf{\Pi}\tilde{\mathbf{s}}' & \mathbf{\Pi}\,\mathbf{L}_t \\ 0 & \mathbf{\Pi} & \mathbf{\Pi}\,\mathbf{L}_r \end{bmatrix} \left\{ \begin{array}{c} \ddot{\mathbf{R}} \\ \ddot{\boldsymbol{\theta}} \\ \ddot{\mathbf{s}}_n' \end{array} \right\} = - \left\{ \begin{array}{c} \ddot{\mathbf{r}}_k'^k \\ \dot{\boldsymbol{\omega}}_k^k \end{array} \right\} - \mathbf{E}^k \left\{ \begin{array}{c} \ddot{\mathbf{\Pi}}^{k\,T}\mathbf{r} + 2\dot{\mathbf{\Pi}}^{k\,T}\dot{\mathbf{r}} + \mathbf{\Pi}^{k\,T}(\Omega) \\ \dot{\mathbf{\Pi}}^{k\,T}\mathbf{\Pi}(\boldsymbol{\omega} + \dot{\boldsymbol{\alpha}}) + \mathbf{\Pi}^{k\,T}\mathbf{\Pi}\,\tilde{\boldsymbol{\omega}}\,\dot{\boldsymbol{\alpha}} \end{array} \right\} \tag{70}$$

with $\Omega = 2\mathbf{\Pi}\tilde{\boldsymbol{\omega}}\,\dot{\mathbf{s}}_n' + \mathbf{\Pi}\,\tilde{\boldsymbol{\omega}}\tilde{\boldsymbol{\omega}}\mathbf{s}'$.

Together, the topological and terminal equations form a necessary and sufficient set of motion equations for FMS. The variational equations of motion for FMS may be obtained from the tree joint element which connects the whole system of bodies to the ground body through a path consisting entirely of branches. Substituting the general form of virtual work terminal equations for each element in the cutset equation for this tree body constraint or joint, one gets a complete set of equations of motion for the FMS. In general, the branch coordinates are not independent, but are related by the kinematic constraint equations. Thus, these constraint acceleration equations (70) must be appended to the set of dynamic equations (67), giving all the equations to solve for branch coordinates and finally yields the classical system differential-algebraic equations of motion for FMS.

5. Examples

By exploiting GT methods and virtual work principles, this formulation has been implemented into a computer program called **FlexNet** (for FLEXible NETwork). Given only a spanning tree with the terminal expressions for deformable bodies in the system, this program automatically generates the equations of motion. Since the selection of a proper tree only requires an elementary knowledge of graph theory, the objective consists in choosing an optimal joint tree to keep the number of branch coordinates and constraint equations to a minimum. Since the equations of motion for deformable bodies are function of a good discretization, all the constant tensors that are function of finite elements have been identified and generated by a finite element preprocessor. Hence, the preprocessor generates look-up tables that can be exploited when needed during the dynamic simulation of FMS. To enforce the constraints at the position and velocity levels, an energy algorithm proposed by (Bauchau et al., 1995) has also been implemented.

A similar symbolic software that fully exploits this graph-theoretic approach in multi-domain modelling is MapleSim, commercialized by Maplesoft. This GT formulation has been extended in the second version of the software (MapleSimII, 2011) to the analysis of mechatronic and other multidisciplinary systems such as mechanical, electrical, thermal, signal/control and hydraulic systems which can all be naturally combined in a model diagram similar the one presented in section 2. However, this software is limited to the simulation of rigid bodies and flexible beams only.

5.1 Flexible four-bar mechanism

Let us consider the example of the planar flexible four-bar mechanism described earlier in section 2. For this FMS, shown in figure 2(a), a proper tree has been highlighted in bold in figure 2(b). This example has been previously analyzed by (Khulief, 1992) and was also analyzed with the software MapleSimII. The rigid crank OA has a length of 0.3 m and is driven at a constant angular speed of $\omega = 210\,rad/s$. The flexible coupler link AB and the flexible follower BC are discretized by eight linear 3-D beam elements with bending moments of inertia equal to $3.112 \times 10^{-8}\,m^4$ and have cross-sectional areas of $1.767 \times 10^{-4}\,m^2$. Flexible links AB and BC are made of steel with modulus of elasticity of $2.0 \times 10^{11}\,N/m^2$ with mass densities of $7800\,kg/m^3$. The rigid crank is assumed horizontal at the initial state. The deflection at the mid-point of the links are mesured perpendicular to the initial links. The first two vibration modes are considered in this simulation. Shown in the figures 6(a) and 6(b) are the numerical results for the relative deflection of the mid-point of the links plotted against the crank angle θ_{31}.

Once the topology and parameters for the FMS has been defined, the translational and rotational graphs can be generated automatically. Due to the systematic nature of GT methods, the previous formulation was encoded with relative ease into a general computer program. Exploiting conventional GT methods, the cutset and circuit equations are automatically generated from the given topology. The terminal equations developped in the preceeding section are contained in a library of modelling components that can be easily updated to include new components. From the projected cutset and circuit equations, the set of motion equations governing the dynamic response of a given FMS is automatically assembled that provides insight into its structural motion. The results are in good agreement with those obtained by (Khulief, 1992) and the software (MapleSimII, 2011).

5.2 Deployment of two flexible panels

As a final example, consider the unfolding of the spatial structure, drawn in figure 7(a), composed of two flexible panels (1) and (2) attached on a rigid base. The linear graph of this flexible structure is shown in figure 7(b). Each panel of dimension $2m \times 0.003m \times 2m$ is made of steel with a Young's modulus of $2.1 \times 10^{11}\,N/m^2$, Poisson's coefficient of 0.3 and a mass density of $7800\,kg/m^3$.

The edges comprising the tree are traced in bold. The deformable plates are represented by edges e_{51} and e_{52}. The revolute joint e_{31} connects panel (1) to the ground and revolute joint e_{32} connects panel (2) to panel (1). Each panel is discretized by 32 triangular elements. The flexible panel elements are represented by edges e_{11} and e_{12}. By imposing a symmetrical meshing throughout the panels, this avoids the generation of torsion deformations outside of their respective plane. To assure good alignment of the nodes on which are fixed the revolute joints, rigid beam elements are conveniently pasted along the edges joining these nodes.

The complete deployment of the flexible panels has been achieved in $T = 14$ seconds where angles ψ_1 and ψ_2 goes from 0° to 90° and from 180° to 0°, respectively. The imposed drivers at the articulations of the structure have zero velocity and zero acceleration at the beginning

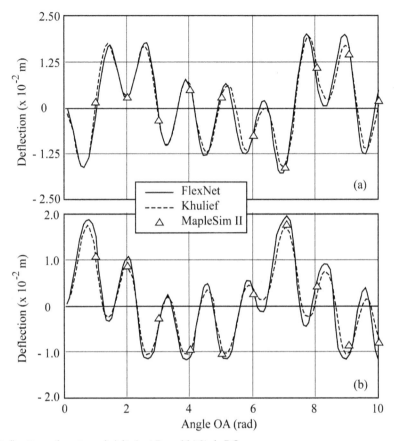

Fig. 6. Deflection of center of a) link AB and b) link BC

and end of the simulation. The angular drivers e_{61} and e_{62} are given by,

$$\psi_1(t) = \frac{\pi}{2T} \left[t - \frac{T}{2\pi} \sin(\frac{2\pi t}{T}) \right] \tag{71}$$

$$\psi_2(t) = \pi - \frac{\pi}{T} \left[t - \frac{T}{2\pi} \sin(\frac{2\pi t}{T}) \right] \tag{72}$$

where $\psi_1(t) = \pi/2$ and $\psi_2(t) = 0$ when $t \geq T$.

Simulations for the flexible panels were performed during $t = 20$ seconds. Results are plotted from the time of release of the panels (at $t = 0s$) through complete deployment (at $t = 14s$) and rebound effects of the panels (until $t = 20s$). Since MapleSimII is restricted to the simulation of flexible links only, figure 8 provides a comparison with the software (Adams/flex, 2011) for the deflection of the centers of the two panels considering the first four vibration modes.

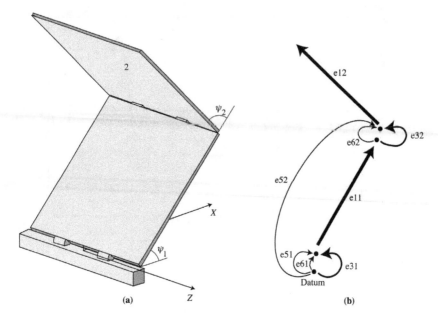

Fig. 7. a) Deployment of the panels and b) graph representation of the system

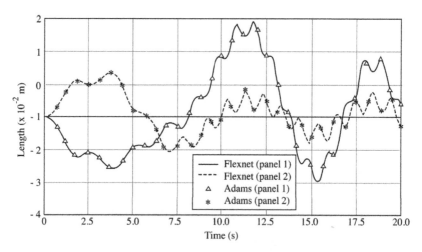

Fig. 8. Transversal deflexion of panel 1 and 2 centers

6. Conclusion

By combining the mechanical system topology with the variational virtual work constitutive equations, a new systematic graph-theoretic formulation has been introduced and used to describe the time-varying configuration of spatial FMS. This method assembles automatically the governing equations of motion in a symmetrical format where the structure and organization of the mass matrix parallels that of structural finite element mass and stiffness matrices, which are also derived using variational methods.

For open-loop systems, like the deployment of panels, a joint tree results in independent branch coordinates and a set of reduced ordinary differential equations can be generated. However, if the graph of a multibody system has closed loops, like in the case of the four-bar mechanism, a tree structure is formed by mathematically cutting the constraining elements or joints yielding the constraint circuit equations. A kinematic formulation may then be developed for the resulting spanning tree, so it neglects momentarily the effect of kinematic constraints between other bodies. While the variational equation of motion is still valid, it holds only for branch coordinate variations that are consistent with the constraint. It is, therefore, necessary to introduce the equation associated with the physical constraint.

Through graph-theoretic methods, the state-of-the-art of general multibody programs has advanced to the point where the flexibility of bodies combined to multi-domain systems can be simulated. Perhaps the most important requirement of a GT general purpose multibody computer program is the quality of the interfaces for the input and output of data. Hence, the self-formulating aspect of all programs exploiting GT methods will become very important for a productive utilization. Other features of GT methods worth mentioning for the future are the portability of the GT algorithms which should be able to adapt easily to all computer environments and the compatibility between different databases of multidisciplinary programs.

7. Acknowledgement

Financial support of this research by the Natural Sciences and Engineering Research Council of Canada is gratefully acknowledged.

8. References

Adams/flex (2011). *MSC/Software Simulating Reality, Delivering Certainty*, http://www.mscsoftware.com/Products/Modeling-Solutions/Default.aspx, last consulted July 2011.

Andrews, G. (1971). *The Vector-Network Model: A topological Approach to Mechanics*, Ph.D. Thesis, University of Waterloo, Waterloo, Ontario, Canada.

Andrews, G. (1977). *A General Re-statement of the Laws of Dynamics Based on Graph Theory*, Problem Analysis in Science and Engineering, Academic Press.

Andrews, G. & Kesavan, H. (1975). The vector-network model: A new approach to vector dynamics, *Mechanism and Machine Theory* Vol. 10: 57–80.

Arczewski, K. (1990). Application of graph theory to the mathematical modelling of a class of rigid body systems, *Journal of the Franklin Institute* Vol. 327, no. 2: 209–220.

Baciu, G., Chou, J. & Kesavan, H. (1990). Constrained multibody systems: Graph-theoretic newton-euler formulation, *IEEE Transactions on Systems, Man and Cybernetics* Vol. 20, no. 5: 1025–1039.

Bauchau, O., Damilano, G. & Theron, N. (1995). Numerical integration of nonlinear elastic multi-body systems, *International Journal of Numerical Methods in Engineering* Vol. 38: 2727–2751.

Behzad, M. & Chartrand, G. (1971). *Introduction to the theory of graphs*, Allyn and Bacon.

Bos, A. (1986). *Modelling Multibody Systems in Terms of Multibond Graphs with Application to a Motorcycle*, Dissertation Twente University.

Chou, J., Kesavan, H. & Singhal, K. (1986). Dynamics of 3-d isolated rigid-body systems: Graph-theoretic models, *Mechanism and Machine Theory* Vol. 21, no. 3: 261–281.

Christensen, R. (1975). *Theory of Viscoelasticity: An Introduction*, Academic Press, New-York.

Christofides, N. (1975). *Graph theory, An Algorithmic Approach*, Academic Press, New York.

Even, S. (1979). *Graph Algorithms*, Computer Science Press.

Flugge, W. (1967). *Viscoelasticity*, Blaisdell, New-York.

Hu, Y. (1988). Applications of bond graphs and vector bond graphs to rigid body dynamics, *Journal of China Textile University* Vol. 5, no. 4: 67–80.

Khulief, Y. (1992). On the finite element dynamic analysis of flexible mechansims, *Computer Methods in Applied Mechanics and Engineering* Vol. 97: 23–32.

Koenig, H. & Blackwell, W. (1960). Linear graph theory - a fundamental engineering discipline, *IRE Transaction on Education* Vol. 3: 42–62.

Koenig, H., Tokad, Y. & Kesavan, H. (1967). *Analysis of Discrete Physical Systems*, McGraw-Hill.

MapleSimII (2011). *High-Performance Multi-Domain Modeling and Simulation*, http://www.maplesoft.com/products/maplesim/index.aspx, last consulted July 2011.

McPhee, J. (1998). Automatic generation of motion equations for planar mechanical systems using the new set of branch coordinates, *Mechanism and Machine Theory* Vol. 33, no. 6: 805–823.

McPhee, J. & Redmond, S. (2006). Modelling multibody systems with indirect coordinates, *Computer Methods in Applied Mechanics and Engineering* Vol. 195, no. 50: 6942–6957.

Richard, M. (1985). *Dynamic Simulation of Constrained Three Dimensional Multibody Systems Using Vector Network Techniques*, Ph.D. Thesis, Queen's University, Kingston, Ontario, Canada.

Richard, M., Bouazara, M. & Therien, J. (2011). Analysis of multibody systems with flexible plates using variational graph- theoretic methods, *Multibody System Dynamics* Vol. 25: 43–63.

Richard, M., Huang, M. & Bouazara, M. (2004). Computer aided analysis and optimal design of mechanical systems using vector-network techniques, *Journal of Applied Mathematics and Computation* Vol. 157: 175–200.

Roberson, R. (1984). The path matrix of a graph, its construction and its use in evaluating certain products, *Journal of Computer Methods in Applied Mechanics and Engineering* Vol. 42: 47–57.

Shabana, A. (1986). Transient analysis of flexible multi-body systems - part1: Dynamics of flexible bodies, *Computer Methods in Applied Mechanics and Engineering* Vol. 54: 75–91.

Shi, P. & McPhee, J. (1997). On the use of virtual work in a graph-theoretic formulation for multibody dynamics, *ASME Design Engineering Technical Conference* DETC97/vib-4199.

Shi, P. & McPhee, J. (2000). Dynamics of flexible multibody systems using virtual work and linear graph theory, *Multibody System Dynamics* Vol. 4: 355–381.

Shi, P., McPhee, J. & Heppler, G. (2001). A deformation field for euler-bernoulli beams with applications to flexible multibody dynamics, *Multibody System Dynamics* Vol. 5: 79–104.

Tennich, M. (1994). *Dynamique de systèmes multi-corps flexibles, une approche générale*, Ph.D. Dissertation, Laval University, Québec, Québec.

Wasfy, T. & Noor, A. (2003). Computational strategies for flexible multibody systems, *ASME Applied Mechanics Reviews* Vol. 56, no. 6: 553–613.

Wittenburg, J. (1977). *Dynamics of Systems of Rigid bodies*, Teubner, Stuttgart.

4

Spectral Clustering and Its Application in Machine Failure Prognosis

Weihua Li[1,2], Yan Chen[2], Wen Liu[1] and Jay Lee[2]
[1]School of Mech. & Auto. Eng,
Souch China University of Technology,
[2]NSFI/UCRC Center for Intelligent Maintenance System,
University of Cincinnati,
[1]China
[2]USA

1. Introduction

Machine fault prognosis and health management has received intensive studies for several decades, and various approaches have been taken, such as statistical signal processing, time-frequency analysis, wavelet, and neural networks. Among of them, pattern recognition method provides a systematic approach to acquiring knowledge from fault samples. In fact, mechanical fault diagnosis is essentially a problem of pattern classification.

Many pattern recognition methods have been studied and applied in machine condition monitoring and fault prognosis. Campbell proposed a linear programming approach to engine failure detection (Campbell&Bennett, 2001). In Ypma's study, different learning methods, such as Independent Component Analysis, Self Organising Map, and Hidden Markov Models, were applied in fault feature extraction, novelty detection and dynamic fault recognition (Ypma, 2001). Ge et.al (2004)proposed a support vector machine based method for sheet metal stamping monitoring. Harkat et.al(2007) applied non-linear principal component analysis in sensor fault detection and isolation. Lei and Zuo (2009) implemented the Weighted k Nearest Neighbour algorithm to identify the gear crack level.

However, the information of machine incipient fault is always weak and contaminated by strong noises, and there is always lack of fault samples to train the learning machine. Therefore, the key issue is how to select sensitive features from the dataset for machine incipient faults prognosis, which is related to feature selection and dimension reduction, and is very useful for fault classification.

In most of medical and clinic applications, when the dimensionality of the data is high, for reducing computation complexity, some techniques might be used to project or embed the data into a lower dimensional space while retaining as much information as possible. Classical linear examples are Principal Component Analysis (PCA) (Jolliffe.2002) and Multi-Dimensional Scaling (MDS) (T. F. Cox & M. A. Cox, 2001). The coordinates of the data points in the lower dimension space might be used as features or simply a mean to visualize the data.

However, for common PHM(Prognostic and Health Management) applications, the dimensionality of the data is not as high as those in medical research, and the mapping techniques are mainly applied to reveal the correlation of features as to increase the accuracy of fault detection and identification. The selection of features also can avoid unnecessary sensors used in machine monitoring, considering the high cost maintaining. Nomikos and MacGregor(1994) firstly presented a PCA approach for monitoring batch process, the history information was linear projected onto a low-dimensional space that summarized the key characteristics of normal behaviour by both variable and their time histories. Considering that minor component discarded in PCA might contain important information on nonlinearities, a large amount of nonlinear methods were presented for the process monitoring and chemical process modelling (Dong & McAvoy,1996; Kaspar & Ray,1992; Sang et. al,2005), such as Kernal PCA (Schölkopf,1998).

Non-linear dimensionality mapping methods are more frequently recognized as non-linear manifold learning methods. The manifold learning is the process of estimating a low-dimensional underlying structure embedded in a collection of high-dimensional data(Tenenbaum et. al, 2000; Roweis & Saul, 2000). Instead of using Euclidian distance to measure samples' similarity in input space, samples' similarity in latent space is measured by their geodesic or short path distance. The deceptive close distance in the high-dimensional input space can be corrected.

Spectral clustering is a graph-theory-based manifold learning method, which can be used to dissect the graph and get the clusters for exploratory data analysis. Compared with the traditional algorithms such as k-means, spectral clustering has many fundamental advantages. It is more flexible, capturing a wider range of geometries, and it is very simple to implement and can be solved efficiently by standard linear algebra methods. It has been successfully deployed in numerous applications in areas such as computer vision, speech recognition, and robotics. Moreover, there is a substantial theoretical literature supporting spectral clustering (Kannan et.al,2004; Luxburg,2007,2008).

In most PHM applications, multi-groups of data sets from different failure modes are frequently nonlinearly distributed and mixed in a high dimensional feature space. However, an "unfolded" feature space is expected as to differentiate these degradation patterns by a designed classifier.

In this part, we first propose a spectral clustering based feature selection method used for machine fault feature extraction and evaluation, and then the samples with selected features are input into a density-adjustable spectral kernel based transductive support vector machine to train and to get the prognosis results.

2. Spectral clustering feature selection

2.1 Basics of graph theory

Given a d-dimentsional data points $\{x_1, . . ., x_n\}$, and the similarity between all pairs of data points x_i and x_j is noted as w_{ij}. According to graph theory, the data points can be represented by an undirected data graph G=(V,E). Each node in this graph represents a data point x_i. Two nodes are connected if the similarity w_{ij} between the corresponding data x_i and x_j is positive or larger than a certain threshold, and the edge is weighted by w_{ij}. These data points can be divided into several groups such that points in the same group are similar and points in different groups are dissimilar to each other.

2.2 Laplacian embedding

BelKin(2003) indicated that Laplacian Eigenmaps used spectral techniques to perform dimensionality reduction. This technique relies on the basic assumption that the data lies in a low dimensional manifold in a high dimensional space. The Laplacian of the graph obtained from the data points may be viewed as an approximation to the Laplace-Beltrami operator defined on the manifold. The embedding maps for the data come from approximations to a natural map that is defined on the entire manifold.

The popular Laplacian Embedding algorithm includes the following steps, as shown in Fig.1.

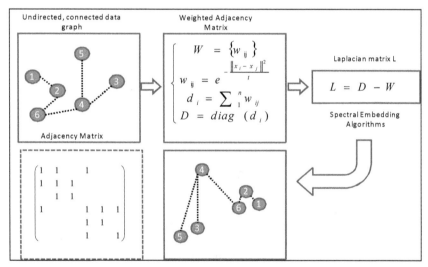

Fig. 1. The procedure of Laplacian Embedding Algorithm

Step 1: The d-dimensional dataset is viewed as an undirected data graph [10] , G = (V, E) with node set V={$x_1,...,x_n$}. Every node in the graph is one point in $\Re_d$. An edge is used to link node i and node j, if they are close as ε–neighborhoods which means the distance between nodes $\mathbf{X}_i$ and $\mathbf{X}_j$ satisfying $\left\| \mathbf{X}_i - \mathbf{X}_j \right\| < \varepsilon$, or if node $\mathbf{X}_i$ is among n nearest neighbors of $\mathbf{X}_j$ or $\mathbf{X}_j$ is among n nearest neighbors of $\mathbf{X}_i$.

Step 2: Each edge between two nodes $\mathbf{X}_i$ and $\mathbf{X}_j$ carries a non-negative weight $w_{ij} \geq 0$. The weighted adjacency matrix of the graph is the matrix $\mathbf{W} = \{w_{ij}\}, i, j = 1,...,n$. There are different methods to configure the weight matrix. For example, the most common is

$$w_{ij} = \begin{cases} 1 & \text{if } x_i \text{ and } x_j \text{ is connected} \\ 0 & \text{otherwise} \end{cases} \tag{1}$$

Or Heat kernel

$$w_{ij} = \begin{cases} e^{-\|x_i - x_j\|^2 / t} & \text{if } x_i \text{ and } x_j \text{ is connected} \\ 0 & \text{otherwise} \end{cases} \tag{2}$$

The degree of a node $X_i \in V$ is defined as $d_i = \sum_1^n w_{ij}$. The degree matrix D is defined as the diagonal matrix with $\{d_1, d_2,...,d_n\}$ on its diagonal. The un-normalized graph Laplacian matrix is defined by Luxburg(2007) as: $\mathbf{L = D - W}$.

Step 3: The Laplacian Eigenmap (on normalized Laplacian matrix) is computed by spectral decomposition for eigenvectors problem of $\mathbf{L}y = \lambda \mathbf{D}y$. The image of X_i under the embedding is converted into the lower dimensional space $\Re m$, given by ordered eigenvalues: $\{y_1(i), y_2(i),..., y_m(i)\}$. This decomposition provides significant information about the graph and distribution of all points. It has been proven experimentally that the inner natural groups of dataset are recovered by mapping the original dataset into the space spanned by eigenvectors of the Laplacian matrix(Belkin & Niyogi,2003).

2.3 Supervised feature selection criterion by Laplacian scores

Given a graph G, the Laplacian matrix L of G is a linear operator on any feature vector from $\mathbf{f} = \{\mathbf{f}_1,...,\mathbf{f}_m\}, \mathbf{f}_i \in R^n$

$$\mathbf{f}_k^T \mathbf{L} \mathbf{f}_k = \frac{1}{2} \sum_{i,j=1}^n w_{ij}(x_{ki} - x_{kj})^2 \tag{3}$$

The equation quantifies how much the feature vector is consistent with the structure of the G locally. For the instances closer to each other, the features that have similar value for them are contributes more on the dissimilarity matrix that is consistent with data structure. The flatter the feature value is over all instances, the smaller the value of the equation. However, instead of the feature consistency only considering instances with small distance, a complete definition of feature consistency with the data structure is clarified as the following:

Definition 1: (feature local consistency with data graph)

Given data graph G= $(\mathbf{V}, \mathbf{E})$ ($V=\{X_1,...,X_n\},E=\{W_{ij}\}$), the feature $\mathbf{f}$ is a locally consistent variant of G at level h ($0<h<1$) for a clustering C over G. If for every cluster C_k of C, there is

$$\frac{\frac{1}{n_k} \sum_{i,j \in C_k} (\mathbf{f}_i - \mathbf{f}_j)^2}{\frac{1}{n_k(n - n_k)} \sum_{i \in C_k, j \notin C_k} (\mathbf{f}_i - \mathbf{f}_j)^2} = h_k \tag{4}$$

And $h = \max(h_k)$ is defined as feature consistency index.

The definition is a ratio between inner and intra cluster variation caused by the individual feature. Perfect clustering expects less variance inter-cluster and the inverse for intra-clusters. If the feature f contributes to better clustering, the nominator tends to be smaller and denominator is larger. Therefore h_k is expected to be smaller. The feature consistency index h indicates the features's weakest separablility for clustering C

In terms of graph theory, similar criterion can be formulated based on Eq.(4), and configure data graph G with following similarity measurement,

$$w^{(1)}_{ij} = \begin{cases} 1/n_k & i,j \in C_k \\ 0 & \text{otherwise} \end{cases} \tag{5}$$

$$w^{(2)}_{ij} = \begin{cases} 0 & i,j \in C_k \\ -1/(n-n_k) & \text{otherwise} \end{cases} \tag{6}$$

Where $w^{(1)}_{ij}$ is the similarity measurement of samples within-class, and $w^{(2)}_{ij}$ that of samples between-class. Then the sequence of instances can be reordered to make the adjacency matrix carry closer instances along its diagonal.

$$\mathbf{W}^{(2)} = \begin{pmatrix} W_1^{(2)} & \cdots & -1/n_1(n-n_1) \\ \vdots & \ddots & \vdots \\ 1/n_p(n-n_p) & \cdots & W_{n_p}^{(2)} \end{pmatrix} \tag{7}$$

$$\mathbf{W}^{(1)} = \begin{pmatrix} W_1^{(1)} & \cdots & 0 \\ \vdots & \ddots & \vdots \\ 0 & \cdots & W_{n_p}^{(1)} \end{pmatrix} \tag{8}$$

As proved by He et.al (2006), Laplacian score of r-th feature is as the follows:

$$L_r = \frac{\tilde{f}_r^T L \tilde{f}_r}{\tilde{f}_r^T D \tilde{f}_r} \tag{9}$$

Because of $\tilde{f}_r^T L \tilde{f}_r = f_r^T L f_r$ (He et.al, 2006), and with the weight matrix as $\mathbf{W}^{(2)}$ and $\mathbf{W}^{(1)}$ as well as their degree diagonal matrixes,

$$\mathbf{D}^{(1)} = -\mathbf{D}^{(2)} = \begin{pmatrix} 1/n_1 & 0 & 0 \\ 0 & \ddots & \vdots \\ 0 & \cdots & 1/n_p \end{pmatrix} \tag{10}$$

The two Laplacian score $\mathbf{L}_r^{(1)}$ and $\mathbf{L}_r^{(2)}$ have same absolute value of denominators. If there exists clustering C={$C_1,\dots, C_p$} over data graph G, the nominators are as the following

$$\mathbf{f}_r^T\mathbf{L}^{(1)}\mathbf{f}_r = \frac{1}{2}\sum_{k=1}^{p}\frac{1}{n_k}\sum_{i,j\in k}(x_{ri}-x_{rj})^2 \tag{11}$$

$$\mathbf{f}_r^T\mathbf{L}^{(?)}\mathbf{f}_r = -\frac{1}{2}\sum_{k=1}^{p}\frac{1}{n_k(n-n_k)}\sum_{i\in k, j\notin k}(x_{ri}-x_{rj})^2 \tag{12}$$

Combining Eq.(4) and Eq.(11), there is

$$\mathbf{f}_r^T\mathbf{L}^{(1)}\mathbf{f}_r = \frac{1}{2}\sum_{k=1}^{p}\frac{h_k}{n_k(n-n_k)}\sum_{i\in k, j\notin k}(x_{ri}-x_{rj})^2 \tag{13}$$

Comparing Eq.(12) with Eq.(13), it can be obtained

$$\min(h_k) < -\frac{\mathbf{f}_r^T\mathbf{L}^{(1)}\mathbf{f}_r}{\mathbf{f}_r^T\mathbf{L}^{(2)}\mathbf{f}_r} = \frac{\mathbf{L}_r^{(1)}}{\mathbf{L}_r^{(2)}} < \max(h_k) = h \tag{14}$$

Therefore, from Eq.(14), instead of the feature consistency index in Definition 1, the ratio of two Laplacian scores can also be considered as equivalent estimation of feature consistency. They are over the data graph with the configuration of $\mathbf{W}^{(2)}$ and $\mathbf{W}^{(1)}$. If the feature is consistent with these data graphs, term of $\mathbf{L}_r^{(1)}$ should be smaller and $\mathbf{L}_r^{(2)}$ be larger.

Therefore, from graph theory perspective, the supervised feature selection criterion by Laplacian score can be defined as follows

$$m = -\frac{\mathbf{f}_r^T\mathbf{L}^{(1)}\mathbf{f}_r}{\mathbf{f}_r^T\mathbf{L}^{(2)}\mathbf{f}_r} = \frac{\mathbf{L}_r^{(1)}}{\mathbf{L}_r^{(2)}} \tag{15}$$

Based on the criterion, the feature can be ranked, and a simple searching engine can be defined to select appropriate number of features from the list.

3. Spectral kernel transductive support vector machine

3.1 Density-adjustable spectral clustering

Commonly, the weight of the edge in a Graph is defined by the Euclid distance between the two nodes, and it works very well with the linear data.

But for nonlinear data, such as two clusters shown in Fig.2, data points a and c belong to the same cluster, and the Euclid distance between points a and b is less than that between points a and c. Therefore, it is necessary to measure the similarity of data points in a different way, which can zoom out the path length of those passing through low density area, and zoom in those not. Then the minimum path can be obtained to replace the Euclid distance. It is very useful for machine failure prognosis, because there always exists nonlinear when machine anomaly occurring. Chapelle et.al (2005) proposed a density-sensitive distance based on a density-adjustable path length definition as follows,

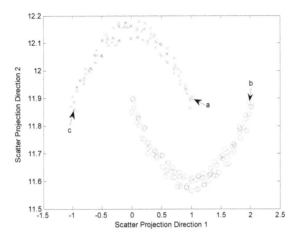

Fig. 2. Scatter Plot of two clusters based on Density-adjustable spectral clustering

$$l(x_i, x_j) = \rho^{dist(x_i, x_j)} - 1 \tag{16}$$

Where $dist(x_i, x_j)$ is the Euclid distance between data x_i and data x_j, and ρ is the density adjustble factor($\rho > 1$). This definition is satisfied with the cluster assumption, and can be used to describe the consistency of data structure by adjusting the factor ρ to zoom out or in the length between the two data points. Therefore, the similarity of the data point x_i and x_j can be expressed as following,

$$s_0(x_i, x_j) = \frac{1}{dsp(l(x_i, x_j)) + 1} \tag{17}$$

Where $dsp(l(x_i, x_j))$ is denoted as the minimum distance between data x_i and x_j, which is the shortest path based on density adjustment.

3.2 Transductive support vector machine

Support vector machine is one of supervised learning methods based on statistical learning theory (Vapnik, 1998). Instead of Empirical Risk Minimization (ERM), Structural Risk Minimization (SRM) is an inductive principle for model selection used for learning from finite training data sets, which enhances the generalization ability of the SVM. The key to SVM is the "kernel tricks", by which the nonlinear map can be realized from low dimensional space to high dimensional space. Therefore, the nonlinear classification task in low dimensional space can be converted to a linear classification, which can be solved by finding a best hyperplane in the high dimensional space.

Considering of 2-class data points, there are many hyperplanes that might classify the data. The best hyperplane is the one that represents the largest margin between the two classes, and the distance from this hyperplane to the nearest data point on each side is maximized.

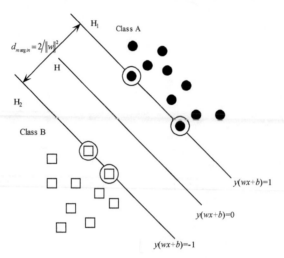

Fig. 3. The Linear Hyperplane of Support Vector Machine

As shown in Fig.3, the data points of Class A are denoted as '•', the others of Class B as '□', and the data points circled by 'o' represented support vectors. These data x in the input space are separated by the best hyperplane H

$$y(\mathbf{w} \cdot x + b) = 0 \tag{18}$$

with the maximal geometric margin

$$\varphi(w) = 2/\|w\|^2 \tag{19}$$

here '·' denotes the dot product and $\mathbf{w}$ is normal vector to the hyperplane, and b is offset from the hyperplane to the margin.

The plane H1 and H2 are also the hyperplanes where the nearest data points to 'H' are located. H1 can be expressed as $y(\mathbf{w} \cdot x + b) = 1$ and H2 $y(\mathbf{w} \cdot x + b) = -1$ respectively. It reveals that finding the best hyperplane means minimizing the $\|w\|^2/2$. There are three widely used kernel function as following,

Polynomial Kernel: $K(x,y) = [\langle x,y \rangle + 1]^d$,

Gaussian Kernel: $K(x,y) = \exp(-\|x - y\|^2/\sigma^2)$,

Hyperbolic: $K(x,y) = \tanh[v(x,y) + c]$.

As for Transductive Support Vector Machine (TSVM), it is one of semi-supervised learning methods, which can combine the labelled data with amounts of unlabelled data co-training. TSVM uses an idea of maximizing separation between labelled and unlabelled data (Vapnik, 1998). It solves

$$\min : \frac{1}{2}\|w\|^2 + C\sum_{i=0}^{l}\xi_i + C^*\sum_{j=0}^{k}\xi_j^* \tag{20}$$

$$s.t.: \forall_{i=1}^{l}: y_i(w \cdot x_i + b) \geq 1 - \xi_i , \; \forall_{j=1}^{k}: y_j(w \cdot x_j + b) \geq 1 - \xi_j^*$$

$$\forall_{i=1}^{l}: \xi_i > 0, \; \forall_{j=1}^{k}: \xi_j^* > 0$$

Where C and C^* are the penalty factors corresponding to labeled and unlabeled data, ξ_i and ξ_j^* are the slack factors respectively, l is the number of labeled data and k that of unlabeled. These parameters are set by user, and they allow trading off margin size against misclassifying training samples or excluding test samples.

3.3 Density-adjustable spectral kernel based TSVM

Combine the ideas of density-adjustable spectral clustering (Chapelle & Zien,2005) and TSVM, we can get the density-adjustable spectral kernel based TSVM algorithm, called DSTSVM. The data is pre-processed by density-adjustable spectral decomposition, and the processed data is input into the TSVM which is trained by gradient descent on a Gaussian kernel, then the data is classified. The implementation of the DSTSVM algorithm is as following,

Input: n-dimension data X{$X_1,...,X_m$} (some labelled and others unlabelled)

Parameter: density-adjustable factor ρ, penalty factor C and kernel width σ of the Gaussian kernel. (Set by user)

Output: The label of unlabelled data and the correctness of classification

Step.1 Calculate the Euclid distance matrix S of data X

Step.2 Calculate the shortest path matrix S_0 according to the Eq.16

Step.3 Construct the Graph G based on data matrix S_0. Define the similarity of between nodes as $w_{ij} = e^{-s_0(x_i,x_j)/2\sigma^2}$, and then the degree diagonal matrix can be denoted as $D(i,i) = \sum w_{ij}$.

Step.4 Calculate the Laplacian matrix $L = D^{(-1/2)}WD^{(-1/2)}$ solve the Eigen-decomposition and rearrange the eigenvalue {$\lambda_1,...,\lambda_n$} and corresponding eigenvector {$U_1,...,U_n$} in descent order.

Step.5 Select the first r nonnegative eigenvectors according to $\left(\sum_{i=1}^{r}\lambda_i / \sum_{j=1}^{n}\lambda_j\right) \geq 85\%$.

Step.6 Get the new data set as $Y = U_r\Lambda_r^{1/2}$ {$y_1,...,y_m$}

Step.7 Train the TSVM by gradient descent using the newdata and then get the classification result.

4. Case study

To demonstrate that the proposed feature selection method and DSTSVM classifier are effective in machine failure prognosis, we applied the methods in feed axis faults feature selection and classification.

4.1 Experiments

Feed axis is one of critical components in a high-precision numerical control machine tool, which always working in conditions such as high speed, heavy duty and large travel distance. This would augment the degradation of mechanical parts such as bearings, ball nuts and so on. From a preventive maintenance perspective, autonomous fault detection and feed axis health assessment could reduce the possibility of causing more severe damage and downtime to machine tool.

TechSolve Inc. collaborated with the NSF Intelligent Maintenance System Center (IMS) to investigate intelligent maintenance techniques for autonomous feed axis failure diagnosis and health assessment. For the investigation, designed experiments were conducted on a feed axis test-bed built by TechSolve. Multiple seeded failures were tested on the system such as axis front and back ball nut misalignment, bearing misalignment and so on (Siegal et.al, 2011). 13 channels (bearing and ball nut accelerometers, temperature and speed; motor power; encode position and so on) data were collected from the test-bed over a period of approximate 6 months. Since all the tests were designed to carry certain failures under different working conditions, the collected information was labeled in terms of the four condition indices including the test index, the load, ball nuts condition, and bearing condition.

Mode	Test index	Table Load	BallNut Misalignment	Bearing Misalignment	Time
1 (Health)	1	0	0	0	2010-10-20
	2	300	0	0	2010-10-22
2 (Failure1)	3	300	0	0.007	2010-11-09
	4	0	0	0.007	2010-11-22
3 (Failure2)	5	0	0.007	0	2010-11-29
	6	300	0.007	0	2010-12-01
4 (Failure1 and Failure2)	7	300	0.007	0.007	2010-12-02
	8	0	0.007	0.007	2010-12-03

Table 1. Eight working conditions of Four modes (Health, Bearing misalignment, Ballnut misalignment, and Combination)

4.2 Feature selection and fault classification

The samples were collected under 4 modes, which were Health, Failure 1(Bearing misalignment $0.007\mu m$), Failure 2 (Ballnut misalignment $0.007\mu m$), and Mode 4 (Failure 1 accompanied with Failure 2). Every mode had two working conditions with load at 0 and 300Kw, and 25 samples at every condition. As for each sample, there were 154 features which contain 117 vibration features (RMS, kurtosis, crest factor at different time periods, and average energy of selected frequency bands) and 37 other features (torque, temperature, position error, and power at different time periods). Therefore, there were totally 200 154-D samples used for investigation.

All the features were evaluated and ranked by Laplacian score using the proposed feature selection criterion. Among 154 features, there were 22 features selected which can reflect the data structure well with the best classification performance, which was shown in Fig.4.

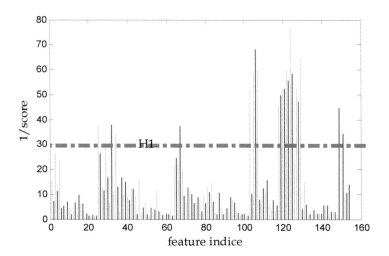

Fig. 4. Features selection based on Laplacian scores

Therefore, the input data dimension can be reduced from 154-D to 22-D. Selecting 25 labelled samples randomly from those 50 22-D samples within every class (totally 100 samples), and the remained 100 samples were regarded as unlabelled ones. Then all these labelled and unlabelled samples were input into the DSTSVM classifier for co-training. This process was repeated for 10 times, and then through 5-fold cross validation, we predicted that which class should the unlabelled samples belong to.

For testing the performance of designed DSTSVM classifier, we reduced the labelled samples to 20 and 10 respectively, and then repeated the procedure above. To verify the effectiveness and correctness, the result was compared with those using SVM (supervised) and TSVM (semi-supervised).

The 10th classification results using the data (10 labelled samples VS 40 unlabelled each class) were shown in Fig.5, Fig. 6, and Fig. 7 respectively.

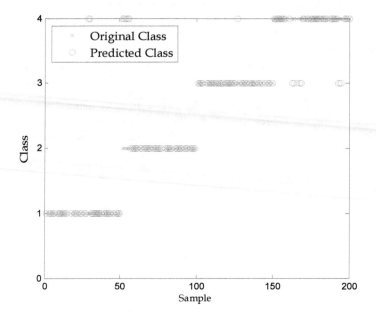

Fig. 5. The Learning result of DSTSVM

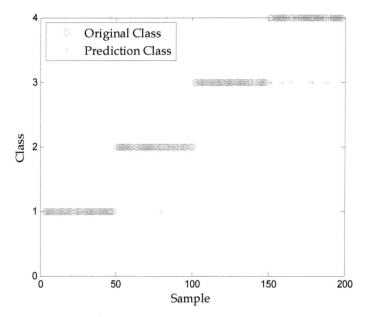

Fig. 6. The Learning result of SVM

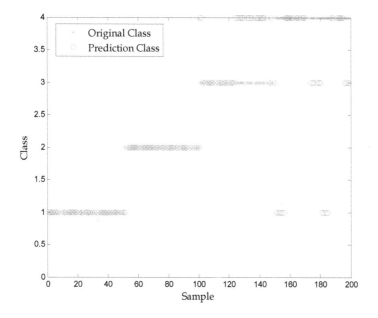

Fig. 7. The Learning result of TSVM

All of the classifiers were trained on Gaussian kernel, and the kernel width σ was set as the optimal value corresponding to the different classifiers. There are two parameters σ and ρ influencing the DSTSVM classification, and the density adjustable factor ρ reflects the data similarity measure, which also affects the Kernel function. In terms of the classification correctness, we can choose the optimal group of these two parameters (ρ,σ). The comparison results under different labelled samples were listed in Table.2.

	DSTSVM		SVM		TSVM	
Labelled vs Unlabelled	Kernel width & density factor (σ,ρ)	Ave Correctness (%)	Kernel width σ	Ave Correctness (%)	Kernel width σ	Ave Correctness (%)
(25vs25)*4	(0.75,2)	91.90	0.5	91.80	0.55	82.50
(20vs30)*4	(0.75,2)	90.92	0.5	90.42	0.55	83.33
(10vs40)*4	(0.75,2)	90.25	0.5	88.88	0.55	85.63

Table 2. The parameters and the average correctness of three classifiers

In Table.2, the average correctness means the average of 10 testing process by 5-fold CV. It can be observed that the proposed method outperforms the TSVM and equals to the supervised SVM under different labelled samples. Moreover, when the labelled data was reduced to 10 samples, it performed better than SVM, which was very meaningful to practical machine failure prognosis applications.

5. Conclusion

The proposed feature selection method can capture the structures of the input data, reduce the dimension of the data and expedite the computation process. More importantly, the classification result is also improved by this feature selection method. Compared with traditional supervised SVM learning and the TSVM semi-supervised learning method, the proposed DSTSVM performed better. Experiment results demonstrate that the proposed DSTSVM method is effective and capable of classifying incipient failures. It has great potential for machine fault prognosis in practice. Based on the current work, the proposed approach can be used to quantify and assure the sufficiency of the data for prognostics applications.

In total, the spectral clustering based method was proposed to evaluate data and to select sensitive features for prognostics, furthermore the spectral kernel based TSVM classifier was also proved to be effective in PHM applications.

6. Acknowledgment

The work described in this paper is supported in part by the National Natural Science Foundation of China (51075150), the Fundamental Research Funds for the Central Universities (2009ZM0091), and Open Research Foundation of State Key Laboratory of Digital Manufacturing Equipment and Technology (DMETKF2009010). The authors would also like to thank Intelligent Maintenance System center in University of Cincinnati and its industry collabrator TechSolve Inc. for the investigation of feed axes system.

7. References

B. Schölkopf, A. Smola, K.R.Muller.(1998). Nonlinear Component Analysis as a Kernel Eigenvalue Problem, *Neural Computation*, Vol.10, No.5,(July 1998), pp.1299–1319, ISSN 0899-7667

Campbell, C. and Bennett, K. (2001). A linear programming approach to novelty detection, *Advances in Neural Information Processing System*, Vol. 14, pp.395-401, ISBN-10 0-262-04208-8, Vancouver, British Columbia, Canada. December 3-8, 2001

Chapelle O, Zien A.(2005). Semi-supervised classification by low density separation, *Proceedings of the 10th International Workshop on Artificial Intelligence and Statistics*, pp57-64, ISBN 0-9727358-1-X, Barbados, January 6-8, 2005

D. Dong and T. J. McAvoy.(1996). Batch tracking via nonlinear principal component analysis, *AIChE Journal*, vol. 42,No.8, (August 1996). pp. 2199-2208, ISSN 1547-5905

Harkat, M-F., Djelel, S., Doghmane, N. and Benouaret, M. (2007). Sensor fault detection, isolation and reconstruction using non-linear principal component analysis, *International Journal of Automation and Computing*, Vol. 4, No. 2, (April 2007),pp.149-155, ISSN 1751-8520

I. Jolliffe.(2002)Principal component analysis, *Encyclopedia of Statistics in Behavioral Science*, ISBN 0-387-95442-2. Springer, New York

J. B. Tenenbaum, V.d.Silva and J.C. Langford.(2000). A global geometric framework for nonlinear dimensionality reduction, *Science*, Vol. 290, No. 5500, (December 2000),pp. 2319-2323, ISSN 0036-8075

T. F. Cox. and M. A. A. Cox (2001). Multidimensional scaling, Chapman and Hall(2nd Edition), ISBN 1-584-88094-3, Florida, USA

M. Belkin and P. Niyogi. (2003). Laplacian Eigenmaps for Dimensionality Reduction and Data Representation, *Neural Computation*,Vol.15,No.6 (June 2003), pp.1373-1396, ISSN 0899-7667

M. Ge, R. Du, G. Zhang, Y. Xu. (2004) Fault diagnosis using support vector machine with an application in sheet metal stamping operations. *Mechanical System and Signal Processing*, Vol.18, No.1, (January 2004), pp.143-159, ISSN 0888-3270

M. H. Kaspar and W. H. Ray.(1992). Chemometric methods for process monitoring and high performance controller design, *AIChE Journal*, Vol. 38, No.10, (October 1992), pp. 1593-1608,ISSN 1547-5905

M. Li, J. Xu, J. Yang, D. Yang and D. Wang. (2009). Multiple manifolds analysis and its application to fault diagnosis, *Mechanical systems and signal processing*, Vol.23, No.9, (November 2009) ,pp. 2500-2509, ISSN 0888-3270

P. Nomikos and J. F. MacGregor.(1994). Monitoring batch processes using multiway principal component analysis, *AIChE Journal*, Vol.40, No.8,(August 1994), pp. 1361-1375, ISSN 1547-5905

Q. Jiang, M.Jia, J. Hu and F. Xu. (2009). Machinery fault diagnosis using supervised manifold learning," *Mechanical systems and signal processing*, Vol.23, No.7, (October 2009), pp. 2301-2311, ISSN 0888-3270

R. Kannan, S. Vempala, and A. Vetta.(2004). On clusterings: Good, bad and spectral. Journal of the ACM, Vol.51, No.3, (May 2004), pp.497–515, ISSN 0004-5411

Skirtich, T., Siegel, D. and Lee, J. (2011). A Systematic Health Monitoring and Fault Identification Methodology for Machine Tool Feed Axis. *MFPT Applied Systems Health Management Conference*, pp.487-506, May 10-12, 2011, Virginia Beach, VA, USA

S. T. Roweis and L. K. Saul.(2000). Nonlinear dimensionality reduction by locally linear embedding, *Science*, Vol. 290, No.5500, (December 2000),pp. 2323-2326, ISSN 0036-8075

U. Von Luxburg. (2007). A tutorial on spectral clustering, *Statistics and Computing*, Vol.17, No.4, (December 2007), pp. 395-416, ISSN 0960-3174

U. von Luxburg, M. Belkin, and O. Bousquet.(2008) Consistency of spectral clustering. *Annals of Statistics*, Vol.36, No.2, (April 2008),pp.555–586, ISSN 0090-5364

Vapnik, V. (1998). Statistical Learning Theory. Wiley, ISBN 0-471-03003-1, New York, USA

W.C Sang, C Lee, J Lee, H Jin, I Lee. (2005). Fault detection and identification of nonlinear processes based on kernel PCA, *Chemometrics and intelligent laboratory systems*, Vol.75, No.1, (Januray 2005), pp.55-67, ISSN 0169-7439

W. Yan and K. F. Goebel. (2005). Feature selection for partial discharge diagnosis, *Proceedings of the 12th SPIE: Health Monitoring and Smart Nondestructive Evaluation of Structural and Biological Systems IV*, Vol.5768, pp. 166-175, ISBN 0-8194-5749-3, March 19 - 22, 2007, San Diego , California,USA

X. He, D. Cai, P. Niyogi.(2006). Laplacian score for feature selection, *Advances in neural information processing systems*, Vol.18, p507-515, ISBN 0-262-19568-2, December 4-7,2006, Vancouver, British Columbia, Canada

Y Lei, M Zuo.(2009). Gear crack level identification based on weighted K nearest neighbor
          classification algorithm, *Mechanical Systems and Signal Processing*, Vol.23,No.5, (July
          2009),pp.1535-1547, ISSN 0888-3270
Ypma, A. (2001). Learning methods for machine vibration analysis and health monitoring,
          PhD Dissertation, ISBN 90-9015310-1, Delft University of Technology, Delft,
          Netherlands.
Z. Zhao and H. Liu, (2007). Spectral feature selection for supervised and unsupervised
          learning, Proceedings of the 24th International Conference on Machine Learning ,
          pp.1151-1157. ISBN 978-1-59593-793-3, June20-24 ,2007, Corvallis, OR, USA

5

Centralities Based Analysis of Complex Networks

Giovanni Scardoni and Carlo Laudanna
Center for BioMedical Computing (CBMC), University of Verona
Italy

1. Introduction

Characterizing, describing, and extracting information from a network is by now one of the main goals of science, since the study of network currently draws the attention of several fields of research, as biology, economics, social science, computer science and so on. The main goal is to analyze networks in order to extract their emergent properties (Bhalla & Iyengar (1999)) and to understand functionality of such complex systems. Two possible analysis approaches can be applied to a complex network: the first based on the study of its topological structure, the second based on the dynamic properties of the system described by the network itself. Since "always structure affects function" (Strogatz (2001)), the topological approach wants to understand networks functionality through the analysis of their structure. For instance, the topological structure of the road network affects critical traffic jam areas, the topology of social networks affects the spread of information and disease and the topology of the power grid affects the robustness and stability of power transmission. Remarkable results have been reached in the topological analysis of networks, concerning the study and characterization of networks structure, and even if far from being complete, several key notions have been introduced. These unifying principles underly the topology of networks belonging to different fields of science. Fundamental are the notions of scale-free network (Barabasi & Albert (1999); Jeong et al. (2000)), cluster (Newman (2006)), network motifs (Milo et al. (2002); Shen-Orr et al. (2002)), small-world property (Watts & Strogatz (1998); Watts (1999); Wagner & Fell (2001)) and centralities. Particularly, centralities have been initially applied to the field of social science (Freeman (1977)) and then to biological networks (Wuchty & Stadler (2003)). Usually, works regarding biological networks rightly consider global properties of the network and when centralities are used, they are often considered from a global point of view, as for example analyzing degree or centralities distribution (Jeong et al. (2000); Wagner & Fell (2001); Wuchty & Stadler (2003); Yamada & Bork (2009); Joy et al. (2005)). A node-oriented approach have been used analyzing attack tolerance of network, where consequences of central nodes deletion are studied (Albert et al. (2000); Crucitti et al. (2004)). But also in this case the analysis have been concentrate on global properties of the network and not on the relevance of the single nodes in the network. Similarly, available software for network analysis is usually oriented to global analysis and characterization of the whole networks. To identify relevant nodes of a biological network, protocols of analysis integrating centralities analysis and lab experimental data are needed and the same for software allowing this kind of analysis. Cytoscape is an excellent visualization and analysis tool with the analysis features greatly enhanced by plug-ins. Plug-in such as NetworkAnalyzer (Assenov et al. (2008)) computes some node centralities but does not allow direct integration with experimental data. Applications such as VisANT

(Hu et al. (2005)), and Centibin (Junker et al. (2006)) calculate centralities, although they either calculate fewer centralities or are not suitable to integration with experimental data. Starting from these general considerations, the first part of this chapter concern the application of network centralities analysis to complex networks from a perspective oriented to identify relevant nodes, with a particular attention to biological networks. Necessary steps to do this are illustrated above through an example of protein-protein interaction network analysis. The aim of the first part of this chapter is to face the centralities analysis of a protein interaction network from a node oriented point of view. The same approach can then be extended to several kinds of complex networks. We want to identify nodes that are relevant for the networks for both centralities analysis and lab experiments. To do this, the following steps have been done:

- Some centralities that we consider significant have been detected. A biological meaning of these centralities have been hypothesized.
- A protocol of analysis for a protein network based on integration of centralities analysis and data from lab experiments (activation level) have been designed.
- A software (CentiScaPe) for computing centralities and integrating topological analysis results with lab experimental data set is presented.
- A human kino-phosphatome network have been extracted from a global human protein interactome data-set, including 11120 nodes and 84776 unique undirected interactions obtained from public data-bases.
- CentiScaPe have been applied to this human kino-phosphatome network and activation level (in threonine and thyrosine) of each protein obtained performeing lab experiments have been related to centrality values.
- Proteins important from both topological analysis and activation level have been easily identified: the attention of successive experiments and analysis should be focused on these proteins.

A further step have been introduced. Once we have identified relevant proteins in a network, we are interested in identifying non-obvious relation between these and other proteins in the network. In any network structure, the role of a node depends, not only on the features of the node itself, but also on the topological structure of the network and on the other nodes features. So even if centralities are node properties, they depend also on other nodes. We know that in a protein network nodes can be added or deleted because of different reasons as for example gene duplication (adding) or gene deletion or drug usage (deleting). More generally, If we delete a relevant node in a complex network, the effects of the deletion have impact not only on the single node and its neighbors, but also on other part of the network. For instance, if you are close friend of an important politician of your town, you have a central role in the social network of the town, and consequently your friends have a central role. But if this politician looses his central role, or if he is completely excluded from the political life of the town, for instance because they put him in prison (this correspond to a deletion on the social network), also you loose your central role in the network and the same for those people related to you. The idea is that the impact of an adding or deletion of a node can be measured through the variation of centrality values of the other nodes in the network. Such notion we introduced have been called "network centralities interference". It allows to identify those nodes that are more sensitive to deletion or adding of a particular node in the network. The Interference Cytoscape plugin have been released to allow this kind of analysis (Scardoni & Laudanna (2011)).

Section 2 consists in a review of some centralities considered important with particular consideration for biological networks. For each centrality a possible biological meaning have been treated and some examples illustrate their significance. Section 3 introduce the CentiScaPe software, the Cytoscape plug-in we implemented for computing network centralities. Main feature of the software is the possibility of integrating experimental data-set with the topological analysis. In CentiScaPe, computed centralities can be easily correlated between each other or with biological parameters derived from the experiments in order to identify the most significant nodes according to both topological and biological properties. In section 4 the protocol of analysis is introduced through an example of analysis of a human kino-phosphatome network. Most relevant kinases and phosphatases according to their centralities values have been extracted from the network and their phosphorylation level in threonine and tyrosine have been obtained through a lab experiment. Centrality values and activation (phosphorylation) levels have been integrated using CentiScaPe and most relevant kinases and phosphatases according to both centrality values and activation levels have been easily identified. Section 5 introduce the Interference software to measure the changes in the topological structure of complex networks.

2. Node centralities: definition and description

In this section, some of the classical network centralities are introduced. For each centrality, we present the mathematical definition, a brief description with some examples, and a possible biological meaning in a protein network. A good and complete description of network centralities can be found in (Koschützki et al. (2005)), where also some algorithms are presented. For many centralities indices it is required that network is connected, i.e. each node is reachable from all the others. If not, some centralities can results in infinity values or some other not properly correct computation. Besides some centralities are not defined for directed graph (except of trivial situation), so we will consider here only connected undirected graph.

Preliminary definitions

Let $G = (N, E)$ an undirected graph, with $n = |N|$ vertexes. $deg(v)$, indicate the degree the vertex. $dist(v, w)$ is the shortest path between v and w. σ_{st} is the number of shortest paths between s and t and $\sigma_{st}(v)$ is the number of shortest paths between s and t passing through the vertex v. Notably:

- Vertex = nodes; edges = arches;
- The "distance" between two nodes, $dist(v, w)$ is the shortest path between the two nodes;
- All calculated scores are computed giving to "higher" values a "positive" meaning, where positive does refer to node proximity to other nodes. Thus, independently on the calculated node centrality, higher scores indicate proximity and lower scores indicate remoteness of a given node v from the other nodes in the graph.

2.1 Degree ($deg(k)$)

Is the simplest topological index, corresponding to the number of nodes adjacent to a given node v, where "adjacent" means directly connected. The nodes directly connected to a given node v are also called "first neighbors" of the given node. Thus, the degree also corresponds to the number of adjacent incident edges. In directed networks we distinguish in-degree, when the edges target the node v, and out-degree, when the edges target the adjacent neighbors of v. Calculation of the degree allows determining the "degree distribution" $P(k)$, which gives the probability that a selected node has exactly k links. $P(k)$ is obtained counting the number of nodes $N(k)$ with $k = 1, 2, 3 \ldots$ links and dividing by the total number of nodes N. Determining the degree distribution allows distinguishing different kind of graphs. For instance, a graph with a peaked degree distribution (Gaussian distribution) indicates that the system has a characteristic degree with no highly connected nodes. This is typical of random, non-natural, networks. By contrast, a power-law degree distribution indicates the presence of few nodes having a very high degree. Nodes with high degree (highly connected) are called "hubs" and hold together several nodes with lower degree. Networks displaying a degree distribution approximating a power-law, $P(k) \approx k^{-\gamma}$, where γ is degree exponent, are called scale-free networks (Barabasi & Albert (1999)). Scale-free networks are mainly dominated by hubs and are intrinsically robust to random attacks but vulnerable to selected alterations (Albert et al. (2000); Jeong et al. (2001)). Scale-free networks are typically natural networks.

In biological terms

The degree allows an immediate evaluation of the regulatory relevance of the node. For instance, in signaling networks, proteins with very high degree are interacting with several other signaling proteins, thus suggesting a central regulatory role, that is they are likely to be regulatory "hubs". For instance, signaling proteins encoded by oncogenes, such as HRAS, SRC or TP53, are hubs. Depending on the nature of the protein, the degree could indicate a central role in amplification (kinases), diversification and turnover (small GTPases), signaling module assembly (docking proteins), gene expression (transcription factors), etc. Signaling networks have typically a scale-free architecture.

2.2 Diameter (Δ_G)

ΔG is the maximal distance (shortest path) amongst all the distances calculated between each couple of vertexes in the graph G. The diameter indicates how much distant are the two most distant nodes. It can be a first and simple general parameter of graph "compactness", meaning with that the overall proximity between nodes. A "high" graph diameter indicates that the two nodes determining that diameter are very distant, implying little graph compactness. However, it is possible that two nodes are very distant, thus giving a high graph diameter, but several other nodes are not (see figure 1). Therefore, a graph could have high diameter and still being rather compact or have very compact regions. Thus, a high graph diameter can be misleading in term of evaluation of graph compactness. In contrast a "low" graph diameter is much more informative and reliable. Indeed, a low diameter surely indicates that all the nodes are in proximity and the graph is compact. In quantitative terms, "high" and "low" are better defined when compared to the total number of nodes in the graph. Thus, a low diameter of a very big graph (with hundreds of nodes) is much more meaningful in term of compactness than a low diameter of a small graph (with few nodes). Notably, the diameter enables to measure the development of a network in time.

In biological terms

The diameter, and thus the compactness, of a biological network, for instance a protein-signaling network, can be interpreted as the overall easiness of the proteins to communicate and/or influence their reciprocal function. It could be also a sign of functional

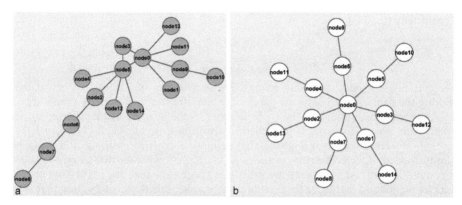

Fig. 1. a. A network where high diameter is due to a low number of nodes. b. A network with low diameter and average distance. The network is "compact".

convergence. Indeed, a big protein network with low diameter may suggest that the proteins within the network had a functional co-evolution. The diameter should be carefully weighted if the graph is not fully connected (that is, there are isolated nodes).

2.3 Average distance (AvD_G)

$$AvD_G = \frac{\sum_{i,j \in N} dist(i,j)}{n(n-1)}$$

where n is the number of nodes in G. The average distance (shortest path) of a graph G, corresponding to the sum of all shortest paths between vertex couples divided for the total number of vertex couples. Often it is not an integer. As for the diameter, it can be a simple and general parameter of graph "compactness", meaning with that the overall tendency of nodes to stay in proximity. Being an average, it can be somehow more informative than the diameter and can be also considered a general indicator of network "navigability". A "high" average distance indicates that the nodes are distant (disperse), implying little graph compactness. In contrast a "low" average distance indicates that all the nodes are in proximity and the graph is compact (figure ??). In quantitative terms, "high" and "low" are better defined when compared to the total number of nodes in the graph. Thus, a low average distance of a very big graph (with hundreds of nodes) is more meaningful in term of compactness than a low average distance of a small graph (with few nodes).

In biological terms

The average distance of a biological network, for instance a protein-signaling network, can be interpreted as the overall easiness of the proteins to communicate and/or influence their reciprocal function. It could be also a sign of functional convergence. Indeed, a big protein network with low average distance may suggest that the proteins within the network have the tendency to generate functional complexes and/or modules (although centrality indexes should be also calculated to support that indication).

2.4 Eccentricity ($C_{ecc}(v)$)

$$C_{ecc}(v) := \frac{1}{max\{dist(v,w) : w \in N\}}$$

The eccentricity is a node centrality index. The eccentricity of a node v is calculated by computing the shortest path between the node v and all other nodes in the graph, then the "longest" shortest path is chosen (let (v, K) where K is the most distant node from v). Once this path with length $dist(v, K)$ is identified, its reciprocal is calculated ($1/dist(v, K)$). By doing that, an eccentricity with higher value assumes a positive meaning in term of node proximity. Indeed, if the eccentricity of the node v is high, this means that all other nodes are in proximity. In contrast, if the eccentricity is low, this means that there is at least one node (and all its neighbors) that is far form node v. Of course, this does not exclude that several other nodes are much closer to node v. Thus, eccentricity is a more meaningful parameter if is high. Notably, "high" and "low" values are more significant when compared to the average eccentricity of the graph G calculated by averaging the eccentricity values of all nodes in the graph.

In biological terms

The eccentricity of a node in a biological network, for instance a protein-signaling network, can be interpreted as the easiness of a protein to be functionally reached by all other proteins in the network. Thus, a protein with high eccentricity, compared to the average eccentricity of the network, will be more easily influenced by the activity of other proteins (the protein is subject to a more stringent or complex regulation) or, conversely could easily influence several other proteins. In contrast, a low eccentricity, compared to the average eccentricity of the network, could indicate a marginal functional role (although this should be also evaluated with other parameters and contextualized to the network annotations).

2.5 Closeness ($C_{clo}(v)$)

$$C_{clo}(v) := \frac{1}{\sum_{w \in N} dist(v,w)}$$

The closeness is a node centrality index. The closeness of a node v is calculated by computing the shortest path between the node v and all other nodes in the graph, and then calculating the sum. Once this value is obtained, its reciprocal is calculated, so higher values assume a positive meaning in term of node proximity. Also here, "high" and "low" values are more meaningful when compared to the average closeness of the graph G calculated by averaging the closeness values of all nodes in the graph. Notably, high values of closeness should indicate that all other nodes are in proximity to node v. In contrast, low values of closeness should indicate that all other nodes are distant from node v. However, a high closeness value can be determined by the presence of few nodes very close to node v, with other much more distant, or by the fact that all nodes are generally very close to v. Likewise, a low closeness value can be determined by the presence of few nodes very distant from node v, with other much closer, or by the fact that all nodes are generally distant from v. Thus, the closeness value should be considered as an "average tendency to node proximity or isolation", not really informative on the specific nature of the individual node couples. The closeness should be always compared to the eccentricity: a node with high eccentricity + high closeness is very

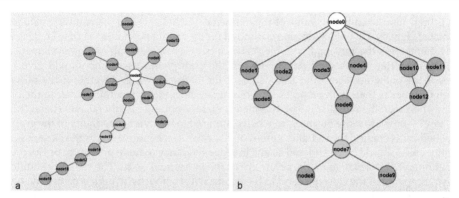

Fig. 2. a. The network shows the difference between eccentricity and closeness. The values of eccentricity are node0=0.14, node8=0.2, node15=0.2. The closeness values are node0=0.021, node8=0.017, node15=0.014. In this case node0 is closer than node8 and node15 to the most of nodes in the graph. Eccentricity value of node0 is smaller than value of node8 and node15, but this is due only to few nodes. If they are proteins this probably mean that node0 is fundamental for the most of reaction in the network, and that node8 and node15 are important only in reactions between few proteins. b. The network shows the difference between centroid and closeness. Here node0 has highest centroid value (centroid=1, closeness=0,04) and node7 has highest closeness value (centroid=-1, closeness= 0,05).

likely to be central in the graph. Figure 2 shows an example of difference between closeness and eccentricity.

In biological terms

The closeness of a node in a biological network, for instance a protein-signaling network, can be interpreted as a measure of the possibility of a protein to be functionally relevant for several other proteins, but with the possibility to be irrelevant for few other proteins. Thus, a protein with high closeness, compared to the average closeness of the network, will be easily central to the regulation of other proteins but with some proteins not influenced by its activity. Notably, in biological networks could be also of interest to analyze proteins with low closeness, compared to the average closeness of the network, as these proteins, although less relevant for that specific network, are possibly behaving as intersecting boundaries with other networks. Accordingly, a signaling network with a very high average closeness is more likely organizing functional units or modules, whereas a signaling network with very low average closeness will behave more likely as an open cluster of proteins connecting different regulatory modules.

2.6 Radiality ($C_{rad}(v)$)

$$C_{rad}(v) := \frac{\sum_{w \in N}(\Delta_G + 1 - dist(v, w))}{n - 1}$$

The radiality is a node centrality index. The radiality of a node v is calculated by computing the shortest path between the node v and all other nodes in the graph. The value of each path is then subtracted by the value of the diameter $+1$ ($\Delta G + 1$) and the resulting values are summated. Finally, the obtained value is divided for the number of nodes -1 ($n - 1$). Basically, as the diameter is the maximal possible distance between nodes, subtracting systematically from the diameter the shortest paths between the node v and its neighbors will give high values if the paths are short and low values if the paths are long. Overall, if the radiality is high this means that, with respect to the diameter, the node is generally closer to the other nodes, whereas, if the radiality is low, this means that the node is peripheral. Also here, "high" and "low" values are more meaningful when compared to the average radiality of the graph G calculated by averaging the radiality values of all nodes in the graph. As for the closeness, the radiality value should be considered as an "average tendency to node proximity or isolation", not definitively informative on the centrality of the individual node. The radiality should be always compared to the closeness and to the eccentricity: a node with high eccentricity + high closeness+ high radiality is a consistent indication of a high central position in the graph.

In biological terms

The radiality of a node in a biological network, for instance a protein-signaling network, can be interpreted as the measure of the possibility of a protein to be functionally relevant for several other proteins, but with the possibility to be irrelevant for few other proteins. Thus, a protein with high radiality, compared to the average radiality of the network, will be easily central to the regulation of other proteins but with some proteins not influenced by its activity. Notably, in biological networks could be also of interest to analyze proteins with low radiality, compared to the average radiality of the network, as these proteins, although less relevant for that specific network, are possibly behaving as intersecting boundaries with other networks. Accordingly, a signaling network with a very high average radiality is more likely organizing functional units or modules, whereas a signaling network with very low average radiality will behave more likely as an open cluster of proteins connecting different regulatory modules. All these interpretations should be accompanied to the contemporary evaluation of eccentricity and closeness.

2.7 Centroid value ($C_{cen}(v)$)

$$C_{cen}(v) := min\{f(v,w) : w \in N\{v\}\}$$

Where $f(v,w) := \gamma_v(w) - \gamma_w(v)$, and $\gamma_v(w)$ is the number of vertex closer to v than to w. The centroid value is the most complex node centrality index. It is computed by focusing the calculus on couples of nodes (v, w) and systematically counting the nodes that are closer (in term of shortest path) to v or to w. The calculus proceeds by comparing the node distance from other nodes with the distance of all other nodes from the others, such that a high centroid value indicates that a node v is much closer to other nodes. Thus, the centroid value provides a centrality index always weighted with the values of all other nodes in the graph. Indeed, the node with the highest centroid value is also the node with the highest number of neighbors (not only first) if compared with all other nodes. In other terms, a node v with the highest centroid value is the node with the highest number of neighbors separated by the shortest path to v. The centroid value suggests that a specific node has a central position within a graph region characterized by a high density of interacting nodes. Also here, "high" and "low" values are more meaningful when compared to the average centrality value of the graph G calculated by averaging the centrality values of all nodes in the graph.

In biological terms

The centroid value of a node in a biological network, for instance a protein-signaling network, can be interpreted as the "probability" of a protein to be functionally capable of organizing discrete protein clusters or modules. Thus, a protein with high centroid value, compared to the average centroid value of the network, will be possibly involved in coordinating the activity of other highly connected proteins, altogether devoted to the regulation of a specific cell activity (for instance, cell adhesion, gene expression, proliferation etc.). Accordingly, a signaling network with a very high average centroid value is more likely organizing functional units or modules, whereas a signaling network with very low average centroid value will behave more likely as an open cluster of proteins connecting different regulatory modules. It can be useful to compare the centroid value to algorithms detecting dense regions in a graph, indicating protein clusters, such as, for instance, MCODE (Bader & Hogue (2003)).

2.8 Stress ($C_{str}(v)$)

$$C_{str}(v) := \sum_{s \neq v \in N} \sum_{t \neq v \in N} \sigma_{st}(v)$$

The stress is a node centrality index. Stress is calculated by measuring the number of shortest paths passing through a node. To calculate the "stress" of a node v, all shortest paths in a graph G are calculated and then the number of shortest paths passing through v is counted. A "stressed" node is a node traversed by a high number of shortest paths. Notably and importantly, a high stress values does not automatically implies that the node v is critical to maintain the connection between nodes whose paths are passing through it. Indeed, it is possible that two nodes are connected by means of other shortest paths not passing through the node v. Also here, "high" and "low" values are more meaningful when compared to the average stress value of the graph G calculated by averaging the stress values of all nodes in the graph.

In biological terms

The stress of a node in a biological network, for instance a protein-signaling network, can indicate the relevance of a protein as functionally capable of holding together communicating nodes. The higher the value the higher the relevance of the protein in connecting regulatory molecules. Due to the nature of this centrality, it is possible that the stress simply indicates a molecule heavily involved in cellular processes but not relevant to maintain the communication between other proteins.

2.9 S.-P. Betweenness ($C_{spb}(v)$)

$$C_{spb}(v) := \sum_{s \neq v \in N} \sum_{t \neq v \in N} \delta_{st}(v)$$

where

$$\delta_{st}(v) := \frac{\sigma_{st}(v)}{\sigma_{st}}$$

The S.-P. Betweenness is a node centrality index. It is similar to the stress but provides a

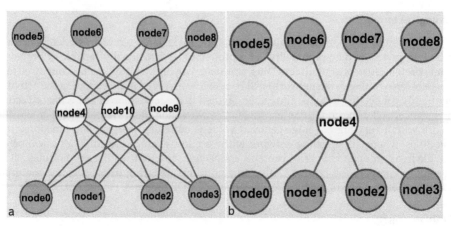

Fig. 3. Betweenness vs Stress. In fig. a node4, node10, and node9 present high value of stress (= 56), and the same value of betweenness (=18.67). In fig.b, node4 presents the same value of stress of fig.a and higher value of betweenness(=56). This is because the number of shortest paths passing through node4 is the same in the two network. But in the second network node4 is the only node connecting the two parts of the network. In this sense betweenness is more precise than stress giving also information on how the node is fundamental in the network. If we remove node4 in fig.a, the connection between the node in the network don't change so much. If we remove node 4 from fig.b the network is completely disconnected.

more elaborated and informative centrality index. The betweenness of a node n is calculated considering couples of nodes $(v1, v2)$ and counting the number of shortest paths linking $v1$ and $v2$ and passing through a node n. Then, the value is related to the total number of shortest paths linking $v1$ and $v2$. Thus, a node can be traversed by only one path linking $v1$ and $v2$, but if this path is the only connecting $v1$ and $v2$ the node n will score a higher betweenness value (in the stress computation would have had a low score). Thus, a high S.-P. Betweenness score means that the node, for certain paths, is crucial to maintain node connections. Notably, to know the number of paths for which the node is critical it is necessary to look at the stress. Thus, stress and S.-P. Betweenness can be used to gain complementary information. Further information could be gained by referring the S.-P. Betweenness to node couples, thus "quantifying" the importance of a node for two connected nodes. Also here, "high" and "low" values are more meaningful when compared to the average S.-P. Betweenness value of the graph G calculated by averaging the S.-P. Betweenness values of all nodes in the graph.

In biological terms

The S.-P. Betweenness of a node in a biological network, for instance a protein-signaling network, can indicate the relevance of a protein as functionally capable of holding together communicating proteins. The higher the value the higher the relevance of the protein as organizing regulatory molecule. The S.-P. Betweenness of a protein effectively indicates the capability of a protein to bring in communication distant proteins. In signaling modules, proteins with high S.-P. Betweenness are likely crucial to maintain functionally and coherence of signaling mechanisms.

2.10 Normalization and relative centralities

Once centralities have been computed, the question that arise immediately is what does it means to have a centrality of, for example, 0.4 for a node? This clearly depends on different parameters as the number of nodes in the network, the maximum value of the centrality and on the topological structure of the network. In order to compare centrality scores between the elements of a graph or between the elements of different graphs some kind of normalization of centrality values is needed. Common normalizations applicable to most centralities are to divide each value by the maximum centrality value or by the sum of all values. We will use the second defining it as the relative centralities value. So, given a centrality C, $C(G,n)$ is the value of the centrality of node n in the network G. We define the relative centrality value of node n as:

$$relC(G,n) = \frac{C(G,n)}{\sum_{i \in N} C(G,i)}$$

So a relative centrality of 0.4 means that the node has the 40% of the total centrality of the network.

2.11 Conclusions

A review of nodes centralities have been presented. The centralities introduced have been chosen for their biological relevance, and a possible biological meaning for each centrality have been hypothesized. Normalization of centralities, useful for comparison between nodes in a network and between nodes of different networks have also been considered.

3. CentiScaPe a software for network centralities

In this section we describe the CentiScaPe software (Scardoni et al. (2009)), a Cytoscape (Cline et al. (2007); Shannon et al. (2003)) plugin we implemented to calculate centralities values and integrating topological analysis of networks with lab experimental data. Main concepts of this section have been published on (Scardoni et al. (2009)).

The vast amount of available experimental data generating annotated gene or protein complex networks has increased the quest for visualization and analysis tools to understand individual node functions masked by the overall network complexity. Cytoscape is an excellent visualization and analysis tool with the analysis features greatly enhanced by plug-ins. Plug-in such as NetworkAnalyzer (Assenov et al. (2008)) computes some node centralities but does not allow direct integration with experimental data. Applications such as VisANT (Hu et al. (2005)), and Centibin (Junker et al. (2006)) calculate centralities, although they either calculate fewer centralities or are not suitable to integration with experimental data. Figure 1 shows a comparative evaluation of CentiScaPe and other applications. CentiScaPe is the only Cytoscape plug-in that computes several centralities at once. In CentiScaPe, computed centralities can be easily correlated between each other or with biological parameters derived from the experiments in order to identify the most significant nodes according to both topological and biological properties. Functional to this capability is the scatter plot by value options, which allows easy correlating node centrality values to experimental data defined by the user. Particularly this feature allows a new way to face the analysis of biological network, integrating topological analysis and lab experimental data. This new approach is described in section 4. At present version 1.2 is available and it is downloaded with a rate of about hundred downloads for month (see Cytoscape

Feature	Centiscape	Network analyzer	Visant	Centibin
Degree	Yes	Yes	Yes	Yes
Radiality	Yes	Yes	No	Yes
Closeness	Yes	Yes	No	Yes
Stress	Yes	Yes	No	Yes
Betweenness	Yes	Yes	No	Yes
Centroid value	Yes	No	No	Yes
Eccentricity	Yes	Yes	No	Yes
Scatter plot between centralities	Yes	No	No	No
Scatter plot with experimental data	Yes	No	No	No
Plot by node	Yes	No	No	No
Highlighting filter	Yes	No	No	No

Table 1. Features of CentiScaPe versus Network Analyzer, Visant, Centibin

website for download statistics). Several results using CentisCaPe have been published in Arsenio Rodriguez (2011); Sengupta et al. (2009a); Lepp et al. (2009); Sengupta et al. (2009b); Biondani et al. (2008); Sengupta et al. (2009c); Feltes et al. (2011); Schokker et al. (2011); Ladha et al. (2010); Choura & Rebaï (2010); Venkatachalam et al. (2011); Webster et al. (2011). **Availability:** CentiScaPe can be downloaded via the Cytoscape web site: http://chianti.ucsd.edu/cyto_web/plugins/index.php. Tutorial, centrality descriptions and example data are available at: http://www.cbmc.it/%7Escardonig/centiscape/centiscape.php

3.1 System overview

CentiScaPe computes several network centralities for undirected networks. Computed parameters are: Average Distance, Diameter, Degree, Stress, Betweenness, Radiality, Closeness, Centroid Value and Eccentricity. Plug-in help and on-line files are provided with definition, description and biological significance for each centrality (see section 2). Min, max and mean values are given for each computed centrality. Multiple networks analysis is also supported. Centrality values appear in the Cytoscape attributes browser, so they can be saved and loaded as normal Cytoscape attributes, thus allowing their visualization with the Cytoscape mapping core features. Once computation is completed, the actual analysis begins, using the graphical interface of CentiScaPe.

3.2 Algorithm and implementation

To calculate all the centralities the computation of the shortest path between each pair of nodes in the graph is needed. The algorithm for the shortest path is the well known Dijkstra algorithm (Dijkstra (1959)). There are no costs in our network edges, so in our case the algorithm keeps one as the cost of each edge. To compute Stress and Betweenness we need all the shortest paths between each pair of nodes and not only a single shortest path between each pair. To do this the Dijkstra algorithm has been adjusted as follows. Exploring the graph when calculating the shortest path between two nodes s and t, the Dijkstra algorithm keep for each node n a predecessor node p. The predecessor node is the node that is the predecessor of n in one of the shortest paths between s and t. So in case of the Dijkstra algorithm, only one predecessor for each node is needed. To have all the shortest paths, we replace the predecessor p with a set of predecessors for each node n. The set of predecessors of the node n is the set of all the predecessors of the node n in the shortest paths set between s and t, i.e. one node is in the set of predecessors of n if it is a predecessor of n in one of the shortest paths between s and t containing n. Once the predecessors set of each node n has been computed, also the tree of all the shortest paths between s and t can be easily computed. Once we have computed all the shortest paths between each pairs of nodes of our network, the algorithm of each centralities comes directly from the formal definition of each centrality. Computational complexity for each centrality value is shown in table 2. A well done description of this and

Centrality	Computational complexity
Diameter	$O(mn + n^2)$
Average distance	$O(mn + n^2)$
Degree ($\deg(v)$)	$O(n)$
Radiality ($\mathrm{rad}(v)$)	$O(mn + n^2)$
Closeness ($\mathrm{clo}(v)$)	$O(mn + n^2)$
Stress ($\mathrm{str}(v)$)	$O(mn + n^2)$
Betweenness ($\mathrm{btw}(v)$)	$O(n^3)$
Centroid Value ($\mathrm{cen}(v)$)	$O(mn + n^2)$
Eccentricity ($\mathrm{ecc}(v)$)	$O(mn + n^2)$

Table 2. Computational complexity for each centrality value. n is the number of nodes and m is the number of edges in the network.

other centralities algorithms can be found in (Koschützki et al. (2005)). CentiScaPe is written in Java as a Cytoscape plugin, in order to exploit all the excellent features of Cytoscape and to reach the larger number of users. The Java library JFreechart (Gilbert (n.d.)) has been used for some graphic features.

3.3 Using CentiScaPe

Once CentiScaPe have been started, the main menu will appear as a panel on the left side of the Cytoscape window as shown in figure 4. The panel shows to the user the list of centralities and the user can select all the centralities or some of them. A banner and a node worked count appear during the computation to show the computation progress. The numerical results are saved as node or network attributes in the Cytoscape attributes browser, depending on the kind of parameters, so all the Cytoscape features for managing attributes are supported: after the computation the centralities are treated as normal Cytoscape attributes. The value of each centrality is saved as an attribute with name "CentiScaPe" followed by the name of the centrality. For example the eccentricity is saved in the Cytoscape attributes browser as "CentiScaPe Eccentricity". Since the Cytoscape attributes follow the alphabetical order this make it easy to find all the centralities in the attributes browser list. There are two kind of centralities: network centralities, and node centralities.

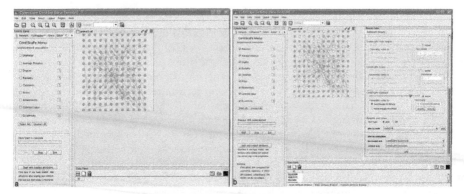

Fig. 4. a. CentiScaPe starting panel. On the left side the main menu appears, to select the centralities for computation. b. A computation results of CentiScaPe. All nodes having centrality values more/equal than the corresponding threshold (AND operator) are highlighted.

Network centralities

The network centralities concern the entire network and not the single nodes. They are the Diameter and the Average Distance. They will appear on the data panel selecting the Cytoscape network attribute browser.

Node centralities

All other centralities are node parameters and refer to the single nodes. So they will appear on the attribute browser as node attributes. Using the Node attribute browser the user can select one or more of them as normal attributes. CentiScaPe also calculates the min, max and mean value for each centrality. Since they are network parameters they appear on the Network attribute browser. As for the other attributes the user can save and load network and node parameters to/from a file. If an attribute is already loaded or calculated and the user try to recalculate it, a warning message will appear.

CentiScaPe results panel

If one or more node centralities have been selected, a result panel will appear on the right side of the Cytoscape window (figure 4b). The first step of the analysis is the Boolean logic-based result panel of CentiScaPe (figure 4b). It is possible, by using the provided sliders in the Results Panel of Cytoscape, to highlight the nodes having centralities values that are higher, minor or equal to a threshold value defined by the user. The slider threshold is initialized to the mean value of each centrality so all the nodes having a centrality value less or equal to the threshold are highlighted by default in the network view with a color depending on the selected visual mapper of Cytoscape (yellow in figure 4b). So if one centrality has been selected, all the nodes having a value less or equal the threshold for that centralities are highlighted. If more than one centralities has been selected they can be joined with an AND or an OR operator. If the AND operator is selected, the nodes for which all the values are less or equal the corresponding threshold are highlighted. If the OR operator is selected the nodes for which at least one value is less or equal the corresponding threshold are highlighted. The possibility of highlighting also the nodes that are more/equal than the threshold is supported. So the user can select the more/equal option for some centralities, the less/equal option for others and can join them with the AND or the OR operator. If necessary, one or more centralities can be deactivated. This feature can immediately answer to questions as: "Which are the nodes having high Betweenness and Stress but low Eccentricity?" Notably, the threshold can also be modified by hand to gain in resolution. In figure ?? are highlighted all the nodes having centralities values more/equal than the corresponding threshold (AND operator). Once the nodes have been selected according to their node-specific values, the corresponding subgraph can be extracted and displayed using normal Cytoscape core features.

Graphic output

Two kind of graphical outputs are supported: plot by centrality and plot by node, both allowing analysis that are not possible with other centralities tools. The user can correlate

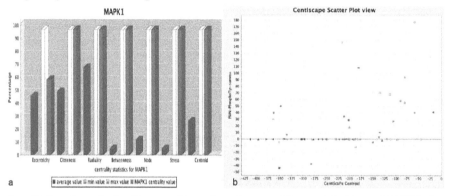

Fig. 5. a. Network analysis of human kino-phosphatome. The protein kinase MAPK1 shows high centralities values for most of the computed centralities suggesting its central role in the network structure and function. For each centrality the specific node value (red), the mean value (blue), the min value (green), and the max value (white) is shown. b. Integration of topological analysis with experimental data. Centroid values are plotted over protein phosphorylation levels in tyrosine. Relevant nodes are easily identified in the top-right quadrant.The centralities values and the node identifier appear in CentiScaPe by passing with the mouse over each geometrical shape in the plot.

centralities between them or with experimental data, such as, for example, gene expression level or protein phosphorylation level (plot by centrality), and can analyze all centralities values node by node (plot by node). Example of plot by node and plot by centrality are shown in figure 5. Graphics can be saved to a jpeg file.

Plot by centrality

The plot by centrality visualization is an easy and convenient way to discriminate nodes and/or group of nodes that are most relevant according to a combination of two selected parameters. It shows correlation between centralities and/or other quantitative node attributes, such as experimental data from genomic and/or proteomic analysis. The result of the plot by centrality option is a chart where each individual node, represented by a geometrical shape, is mapped to a Cartesian axis. In the horizontal and vertical axis, the values of the selected attributes are reported. Most of the relevant nodes are easily identified in the top-right quadrant of the chart. Figure 5b shows a plot of centroid values over intensity of protein tyrosine phosphorylation in the human kino-phosphatome network derived from the analysis of human primary polymorphonuclear neutrophils (PMNs) stimulated with the chemoattractant IL-8 (see section 4). The proteins having high values for both parameters likely play a crucial regulatory role in the network. The user can plot in five different ways: centrality versus centrality, centrality versus experimental data, experimental data versus experimental data, a centrality versus itself and an experimental data versus itself. Notably, a specific way to use the plot function is to visualize the scatter plot of two experimental data attributes. This is an extra function of the plug-in and can be used in the same way of the centrality/centrality option and centrality/experimental attribute option. If the plot by centrality option is used selecting the same centrality (or the same experimental attribute) for both the horizontal and the vertical axis, result is an easy discrimination of nodes having low values from nodes having high values of the selected parameter (figure 6a) Thus, the main use of the "plot by centrality" feature is to identify group of nodes clustered according to combination of specific topological and/or experimental properties, in order to extract sub-networks to be further analyzed. The combination of topological properties with experimental data is useful to allow more meaningful predictions of sub-network function to be experimentally validated.

Plot by node

The plot by node option, another unique feature of CentiScaPe, shows for every single node the value of all calculated centralities represented as a bar graph. The mean, max and min values are represented with different colors. To facilitate the visualization, all the values in the graph are normalized and the real values appear when pointing the mouse over a bar. Figure 5a shows, as an example, the values for the MAPK1 calculated from the global human kino-phosphatome (see section 4).

3.4 Conclusions

CentiScaPe is a versatile and user-friendly bioinformatic tool to integrate centrality-based network analysis with experimental data. CentiScaPe is completely integrated into Cytoscape and the possibility of treating centralities as normal attributes permits to enrich the analysis with the Cytoscape core features and with other Cytoscape plug-ins. The analysis obtained with the Boolean-based result panel, the "plot by node" and the "plot by centrality" options give meaningful results not accessible to other tools and allow easy categorization of nodes in large complex networks derived from experimental data.

4. A new protocol of analysis. Centralities in the human kino-phosphatome

In this section a new protocol of analysis of protein interaction network is introduced through an example of analysis of the human kino-phosphatome (Scardoni et al. (2009)). The analysis starts with the extraction of known interaction from a protein interactome. In our case we consider kinases and phosphatases interaction i.e. those interaction regarding activation and inhibition of proteins in the network. Kinases and phosphatases are enzymes involved in the phosphorilation process: they transfer or remove phosphate groups to/from a protein regulating in this way its activity. Substantially kinases and phosphatases activate or inhibit other proteins in the network. In a kino-phosphatome network this process generates a cascade of activations and inhibitions of proteins corresponding to the transmissions of signals and to the control of complex processes in cells. The approach to the kino-phosphatome network is to identify the most important proteins for their centrality values and then to analyze with a lab experiment their activation level. After this, using the CentiScaPe feature of integrating topological analysis and data from lab experiments, those values are integrated and those nodes important for both centralities value and activation level are easily identified. This introduce a new way of facing the analysis of a protein interaction network based on the Strogatz assertion that in a biological network "Structure always affects function" (Strogatz (2001)). Instead of concentrating the analysis, as usual, on the global properties of the network (such degree distribution, centralities distribution, and so on) we consider in a cause-effect point of view single nodes of the network relating their centrality values (cause) with activation level (effects). Most of the contents of this section have been published on (Scardoni et al. (2009)).

4.1 Centralities analysis

The protocol used for the analysis of the human kino-phosphatome network is the following:

- The nodes of interest are extracted from the global network, resulting in a subnetwork to analyze (in our example the subnetwork of human kinases and phosphatases have been extracted from a human proteins interactome).
- The centralities values are computed with CentiScaPe. A subnetwork of proteins with all centrality values over the average is extracted.
- The lab experiment identifies which of these proteins present high phosphorilation level (in our example in tyrosine and threonine)
- Using CentiScaPe, lab experimental data and centrality values are integrated, so proteins with high level of activation and high centralities values are easily identified.
- Next experiments and analysis should be focused on these proteins.

This protocol have been applied as follows. A global human protein interactome data-set (Global Kino-Phosphatome network), including 11120 nodes and 84776 unique undirected interactions (IDs = HGNC), was complied from public data-bases (HPRD, BIND, DIP, IntAct, MINT, others; see (Scardoni et al. (2009)) on-line file GLOBAL-HGNC.sif) between human protein kinases and phosphatases. The resulting sub-network, a kino-phosphatome network, consisted of 549 nodes and 3844 unique interactions (see (Scardoni et al. (2009)) on-line files Table S4 and Kino-Phosphatome.sif), with 406 kinases and 143 phosphatases. The kino-phosphatome network did not contain isolated nodes. We used CentiScaPe to calculate centrality parameters. A first general overview of the global topological properties of the kino-phosphatome network comes from the min, max and average values of all computed centralities along with the diameter and the average distance of the network (table 3). These data provide a general overview of the global topological properties of the kino-phosphatome network. For instance, an average degree equals to 13.5 with an average distance of 3 may suggest a highly connected network in which proteins are strongly functionally interconnected. Computation of network centralities allowed a first ranking of human kinases and phosphatases according to their central role in the network (see (Scardoni et al. (2009)) on-line files Table S6 reporting all node-by-node values of different centralities). To facilitate the identification of nodes with the highest scores we applied the "plot by centrality" feature of CentiScaPe. A first plotting degree over degree generated a linear distribution, as expected (see fig. 6). However, it is evident that the distribution is

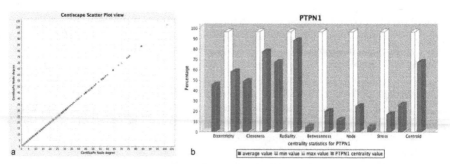

Fig. 6. a. A scatter plot degree over degree. As expected this generate a linear distribution. Notably the distribution is not uniform: many nodes display low degree and only few nodes with high degree, according to the scale-free architecture of biological network. b. The "plot by node" representation for PTPN1. The phosphatase PTPN1 presents the highest degree between all the phosphatases and a rather high score for other centralities. This suggests that PTPN1 may play a central regulatory role in the network. For each centrality the specific node value (red), the mean value (blue), the min value (green), and the max value (white) is shown.

not uniform, with the majority of nodes having a similar low degree and very few having very high degree. This is consistent with the known scale-free architecture of biological networks (Jeong et al. (2000)). The scale-free topology of the kino-phosphatome network was also confirmed with Network Analyzer (Assenov et al. (2008)). A total of 186 nodes (164 kinases and 22 phosphatases) displayed a degree over the average. The top 10 degrees (64 to 102) were all kinases, with MAPK1 showing the highest degree (102). Notably, MAPK1 displayed the highest score for most of the computed centralities (fig. 5a), suggesting its central regulatory role in the kino-phosphatome. In contrast, PTPN1 had the highest degree, 46, between all phosphatases (top 31 among all nodes) and had a rather high score also for other centralities Thus, degree analysis suggests that MAPK1 and PTPN1 are the most central kinase and phosphatase, respectively. To further support this suggestion we analyzed the centroid. Average centroid was -393. 242 nodes (206 kinases and 36 phosphatases) displayed a centroid over the average. The top 10 centroid (-79 to 8) were all kinases, with MAPK1 showing the highest centroid value (18). PTPN1 had the highest centroid value, -154, between all phosphatases (top 22 among all nodes). Thus, as for the degree, also the centroid value analysis suggests a possible scale-free distribution, with MAPK1 and PTPN1 being the most central kinase and phosphatase, respectively. This conclusion is also easily evidenced by plotting the degree over the centroid (fig. 7). Here MAPK1 appears at the top right of the plot and PTPN1 is present in the top most dispersed region of the plot, thus suggesting their higher scores. Interestingly, from the analysis is evident a non-linear distribution of nodes, with few dispersed nodes occupying the top right quadrant of the plot (i.e. high degree and high centroid): these nodes can potentially represent particularly important regulatory kinases and phosphatases. This kind of analysis can be iterated by evaluating all other centralities. To extract the most relevant nodes according to all centrality values we used CentiScaPe to select all nodes having all centrality values over the average. Upon filtering we obtained a kino-phosphatome sub-network (fig.**??**) consisting of 97 nodes (82 kinases and 15 phosphatases) and 962 interactions (see (Scardoni et al. (2009)) on-line files Table S7, and K-P sub-network.sif).

CentiScaPe Average Distance	3.0292037280789224
CentiScaPe Betweenness Max value	20159.799011925716
CentiScaPe Betweenness mean value	1112.0036429872616
CentiScaPe Betweenness min value	0.0
CentiScaPe Centroid Max value	18.0
CentiScaPe Centroid mean value	-393.07285974499086
CentiScaPe Centroid min value	-547.0
CentiScaPe Closeness Max value	8.771929824561404E-4
CentiScaPe Closeness mean value	6.175318530305184E-4
CentiScaPe Closeness min value	3.505082369435682E-4
CentiScaPe Diameter	8.0
CentiScaPe Eccentricity Max value	0.25
CentiScaPe Eccentricity mean value	0.18407494145199213
CentiScaPe Eccentricity min value	0.125
CentiScaPe Radiality Max value	6.91970802919708
CentiScaPe Radiality mean value	5.970796271921072
CentiScaPe Radiality min value	3.7937956204379564
CentiScaPe Stress Max value	210878.0
CentiScaPe Stress mean value	11537.009107468124
CentiScaPe Stress min value	0.0
CentiScaPe degree Max value	102.0
CentiScaPe degree mean value	13.5591985428051
CentiScaPe degree min value	1.0

Table 3. Global values of the kino-phosphatome network computed using CentiScaPe. The table includes min, max and mean value for each centrality and also the global parameter Diameter and Average Distance.

This sub-network possibly represents a group of highly interacting kinases and phosphatases displaying a critical role in the regulation of protein phosphorylation in human cells. Further analysis with CentiScaPe or other analysis tools, such as MCODE (Bader & Hogue (2003)) or Network Analyzer (Assenov et al. (2008)), performing a Gene Ontology database search (Ashburner et al. (2000)), or adding functional annotation data, may allow a deeper functional exploration of this sub network. The regulatory role of proteins belonging to the kino-phosphatome network may be also experimentally tested in a context-selective manner. Indeed, the centrality analysis by CentiScaPe can be even more significant by superimposing experimental data. To test this possibility, we focused the analysis on human polymorphonuclear neutrophils (PMNs).

4.2 Phosphoproteomic analysis of chemoattractant stimulated human PMNs

Human primary polymorphonuclear cells isolation

Human primary polymorphonuclear cells (PMNs) were freshly isolated form whole blood of healthy donors by ficoll gradient sedimentation. Purity of PMN preparation was evaluated by flow cytometry and estimated to about 95% of neutrophils. Isolated PMNs were kept in culture at $37^{\circ}C$ in standard buffer (PBS, $1mM$ $CaCl_2$, $1mM$ $MgCl_2$, 10% FCS, $pH7.2$) and used within 1 hour. Viability before the assays was more than 90%.

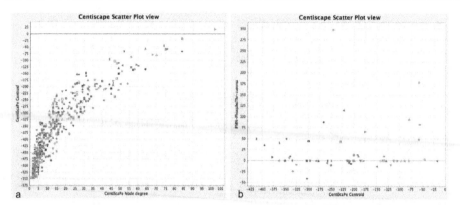

Fig. 7. a. A "plot by centralities" representation of degree over centroid. In the top right of the plot appear the nodes having high values of both degree and centroid (including MAPK1). b. Integration of topological analysis with experimental data. Centroid values are plotted over protein phosphorylation levels in threonine, experimentally determined as described in the text.

Human primary polymorphonuclear cell stimulation

Human neutrophils were resuspended in standard buffer at $10^7/ml$ and stimulated under stirring at 37^oC for 1 min. with the classical chemoattractant fMLP ($100nM$). Stimulation was blocked by directly disrupting the cells for 10 min. in ice-cold lysis buffer containing: $20mM$ MOPS, $pH7.0$, $2mM$ EGTA, $5mM$ EDTA, $30mM$ sodium fluoride, $60mM$ β-glycerophosphate, $20mM$ sodium pyrophosphate, $1mM$ sodium orthovanadate, $1mM$ phenylmethylsulfonylfluoride, $3mM$ benzamidine, $5\mu M$ pepstatin A, $10\mu M$ leupeptin, 1% Triton X-100. Lysates were clarified by centrifugation at $12.000xg$ for 10 min. and kept at 80^oC until further processing.

Evaluation of protein phosphorylation

Protein phosphorylation was evaluated both qualitatively and quantitatively by using the Kinexus protein array service (Kinexus (n.d.)). Kinexus provides a complete service for high throughput proteomic and phosphoproteomic high sensitive analysis of cell lysed samples, allowing detection of more than 800 proteins, including about 200 phosphorylated proteins (about 350 phospho-sites) by means of in-house validated antibody microarrays (Kinexus (n.d.)). $100\mu l$ of frozen samples of lysed PMNs (about $1mg/ml$ protein concentration) have been sent to Kinexus for the analysis. Phosphoproteomic antibody microarray data have been delivered by email and subsequently elaborated to extract values of protein phosphorylation of control versus agonist-triggered samples. (phosphorylation data files are available on-line: see (Scardoni et al. (2009)) PMN-PhosphoSer.NA, PMN-PhosphoTyr.NA, PMN-PhosphoThr.NA).

4.3 Combining topological analysis and experimental data

Data about protein phosphorylation were used as bioinformatic probes and node attributes to extract, from the Global Kino-Phosphatome network, subnetworks of protein phosphorylation, to be analyzed with CentiScaPe Experimental data were loaded as node attributes in Cytoscape and the computed centrality values were plotted over values of protein phosphorylation. Here, every node is represented with two coordinates consisting of a computed centrality and of experimental data regarding protein phosphorylation induced in PMNs by fMLP. In figures 5b and 7b are shown plots of centroid values over intensity of protein phosphorylation in threonine or tyrosine residues induced by fMLP triggering in human PMNs. Notably, in the plot are shown only those proteins whose phosphorylation level was experimentally determined. The two plots allow immediately evidencing that proteins phosphorylated in threonine (fig. 5b) or in tyrosine (fig. 7b) have different topological position in the network, with proteins phosphorylated in tyrosine showing a higher centrality values. This could suggest that tyrosine phosphorylation induced in PMNs by chemoattractants involves signaling proteins regulating clusters of proteins, as the centroid value may suggest. Besides, the top/left quadrant is empty in both figures 5b and 7b. So there are no nodes having low centroid value and high phosphorylation in threonine or tyrosine. This may suggest that centroid value and activation level are strictly related. Further hypotheses can be formulated by expanding the analysis to other centralities and by adding more phopshorylation data. From this type of plotting it is possible to further identify relevant nodes not only according to topological position but also to experimental outputs. Thus, groups of nodes whose regulatory relevance is suggested by centrality analysis are further characterized by the corresponding data of biological activity.

4.4 Conclusions

In this section a protocol of analysis for protein network have been proposed. The key idea is that of identify most important proteins from both topological and biological point of view. Through the example of the kino-phosphatome network, we have seen how CentiScaPe can integrate the two kinds of analysis allowing an easy characterization of most relevant proteins. The topological analysis and experimental data do confirm each other's regulatory relevance and may suggest further, more focused, experimental verifications. Combination of CentiScaPe with other bioinformatics tools may help to analyze high throughput genomic and/or proteomic experimental data and may facilitate the decision process.

5. A further step in centralities analysis: node centralities interference

As seen in the previous section, network centralities allow us to understand the role and the importance of each single node in a protein network. Next step we introduce in this section is to understand and measure changes to the topological structure of the network. The effects of mutation in the network structure, have been studied from a global point of view: nodes are removed from the network and the effects on some global parameters, as for example diameter, average distance or global efficiency are evaluated (Barabasi & Oltvai (2004); Jeong et al. (2001); Albert et al. (2000); Crucitti et al. (2004)). Our approach wants to answer to this question: "we remove or add one node in the network, how do other nodes modify their functionality because of this removal?" Since centrality indexes allow categorizing nodes in complex networks according to their topological relevance (see CentiScaPe plugin), in a node-oriented perspective, centralities are very useful topological parameters to compute in order to quantify the effect of individual node(s) alteration. We introduced the notion of interference and developed the Cytoscape plugin **Interference** (Scardoni & Laudanna (2011)) to evaluate the topological effects of single or multiple nodes removal from a network. In this perspective, interference allows virtual node knock-out experiments: it is possible to remove one or more nodes from a network and analyze the consequences on network structure, by looking to the variations of the node centralities values. As the centrality value of a node is strictly dependent on the network structure and on the properties of other nodes in the network, the consequences of a node deletion are well captured by the variation on the centrality values of all the other nodes. The interference approach can model common situations where real nodes are removed or added from/to a physical network:

- Biological networks, where one or more nodes (genes, proteins, metabolites) are possibly removed from the network because of gene deletion, pharmacological treatment or protein degradation. Interference can be used to:
  - Simulate pharmacological treatment: one can potentially predict side effects of the drug by looking at the topological properties of nodes in a drug-treated network, meaning with that a network in which a drug-targeted node (protein) was removed. To inhibit a protein (for instance a kinases) corresponds to removing the node from the network
  - Simulate gene deletion: gene deletion implies losing encoded proteins, thus resulting in the corresponding removal of one or more nodes from a protein network
- Social and financial networks, where the structure of the network is naturally modified over time
- Power grid failures
- Traffic jam or work in progress in a road network
- Temporary closure of an airport in an airline network

5.1 How Interference plugin works

The Interference plug-in allows you identifying the area of influence of single nodes or group of nodes. Specifically, Interference:

- Compute the centralities value of the network
- Remove the node(s) of interest
- Recompute the centralities in the new network (the one where the node(s) have been removed)
- Evaluate the differences between the centralities in the two networks.

This allows identifying the differences between the two networks: the one with the node(s) of interest still present and the one where the node(s) have been removed. Some nodes will increase, whereas others will reduce, their individual centrality values. This may suggest hints on the functional, regulatory, relevance of specific node(s).

Example

Consider the two networks in figure 8 and observe the role of node1 and node5. Network B is obtained by network A removing node5. Interference notion is based on measuring the effects of such remotion: in the table 4 are reported the betweenness values of the two networks (in percentage). Consider node1: in the network A, node1 has the 27% of the total value of betweenness. In the network B (where node5 has been removed) node1 has the 64% of the total value. The betweenness interference of node5 with respect to node1 is: Betwennes of node1 in the network A - Betwennes of node1 in the network = -37% It means that the topological relevance measured by the betweenness centrality of node1 increase of the 37% of the total betweenness if we remove node5 from the network. The presence of node5 negatively interferes with the central role of node1 measured by the betweennes centrality.

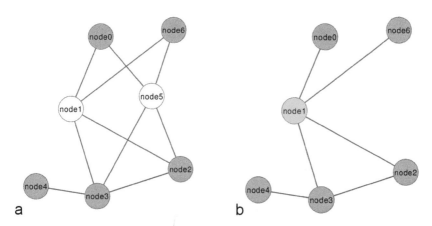

Fig. 8. Network b is obtained by network a removing node5. In this case node1 remains the only node connecting the top of the network with the bottom. Its betweenness values will increase.

Node	Network A (with node5) Betweenness (%)	Network B (no node5) Betweenness (%)	Interference value
node1	27%	64%	-37%
node4	0%	0%	0%
node0	2%	0%	2%
node2	2%	0%	2%
node6	2%	0%	2%
node3	40%	36%	4%
node5	27%		

Table 4. Betweenness and interference values expressed as percentage of the total value for the networks in figure 8. Node1 is the most sensitive to the deletion of node5.

The interference values of node5 with respect to the overall network are reported in the table. Node1 is the node mostly affected by the presence of node5 in the network. If we are considering real networks we expect that the activity of node5 strongly affects the activity of node 1.

Positive interference

If a node (A), upon removal from the network of a specific node (B) or of a group of nodes, decreases its value for a certain centrality index, its interference value is positive. This means that this node (A), topologically speaking, takes advantage (is positively influenced) by the presence in the network of the node (B) or of that group of nodes. Thus, "removal" of node (B) or of that group of nodes from the network, negatively affects the topological role of the node (A). This is called positive interference.

Negative interference

If a node (A), upon removal from the network of a specific node (B) or of a group of nodes, increases its value for a certain centrality index, its interference value is positive. This means that this node (A), topologically speaking, is disadvantaged (is negatively influenced) by the presence in the network of the node (B) or of that group the nodes. Thus, "removal" of node (B) or of that group of nodes from the network, positively affects the topological role of node (A). This is called negative interference.

5.2 Conclusions

As argued above, interference naturally induces cluster of proteins that are similar for their interference values due to the same node. A new clusterization algorithm can be derived if we group nodes depending on their interference value: given a node we compute its interference value and we put all the nodes having high interference in the same cluster. This interference-based modular decomposition of a network characterizes nodes for their answer to the inhibition (or adding) of a certain node in the network. If deletion of the node in a protein network is due to drug usage, the cluster of nodes having high interference value is the set of proteins where the drug has its greatest effects. In pharmacology this should permit to predict which proteins are more affected from the inhibition of another protein in the network. We can so prevent side effects of the inhibition of a node due to a drug usage.

6. References

Albert, R., Jeong, H. & Barabasi, A.-L. (2000). Error and attack tolerance of complex networks, *Nature* 406(6794): 378–382.

Arsenio Rodriguez, D. I. (2011). Characterization in silico of flavonoids biosynthesis in theobroma cacao l., *Network Biology* 1: 34–45.

Ashburner, M., Ball, C. A., Blake, J. A., Botstein, D., Butler, H., Cherry, J. M., Davis, A. P., Dolinski, K., Dwight, S. S., Eppig, J. T., Harris, M. A., Hill, D. P., Issel-Tarver, L., Kasarskis, A., Lewis, S., Matese, J. C., Richardson, J. E., Ringwald, M., Rubin, G. M. & Sherlock, G. (2000). Gene ontology: tool for the unification of biology. the gene ontology consortium., *Nature genetics* 25(1): 25–29.

Assenov, Y., Ramirez, F., Schelhorn, S.-E., Lengauer, T. & Albrecht, M. (2008). Computing topological parameters of biological networks, *Bioinformatics* 24(2): 282–284.

Bader, G. D. & Hogue, C. W. (2003). An automated method for finding molecular complexes in large protein interaction networks., *BMC Bioinformatics* 4(1).

Barabasi, A.-L. & Albert, R. (1999). Emergence of scaling in random networks, *Science* 286(5439): 509–512.

Barabasi, A.-L. & Oltvai, Z. N. (2004). Network biology: understanding the cell's functional organization, *Nature Reviews Genetics* 5(2): 101–113.

Bhalla, U. S. & Iyengar, R. (1999). Emergent properties of networks of biological signaling pathways, *Science* 283.

Biondani, Viollet, Foretz, Laudanna, Devin-Leclerc, Scardoni & Franceschi, D. (2008). Identification of new functional targets of *ampkα*1 in mouse red cells., *48th annual meeting of American society for cell biology*.

Choura, M. & Rebaï, A. (2010). Application of computational approaches to study signalling networks of nuclear and Tyrosine kinase receptors., *Biology direct* 5: 58+.

Cline, M. S., Smoot, M., Cerami, E., Kuchinsky, A., Landys, N., Workman, C., Christmas, R., Avila-Campilo, I., Creech, M., Gross, B., Hanspers, K., Isserlin, R., Kelley, R., Killcoyne, S., Lotia, S., Maere, S., Morris, J., Ono, K., Pavlovic, V., Pico, A. R., Vailaya, A., Wang, P.-L. L., Adler, A., Conklin, B. R., Hood, L., Kuiper, M., Sander, C., Schmulevich, I., Schwikowski, B., Warner, G. J., Ideker, T. & Bader, G. D. (2007). Integration of biological networks and gene expression data using cytoscape., *Nature protocols* 2(10): 2366–2382.

Crucitti, P., Latora, V., Marchiori, M. & Rapisarda, A. (2004). Error and attack tolerance of complex networks, *Physica A: Statistical Mechanics and its Applications* 340(1-3): 388 – 394. News and Expectations in Thermostatistics.

Dijkstra, E. W. (1959). A note on two problems in connexion with graphs, *Numerische Mathematik* 1: 269–271.

Feltes, B., de Faria Poloni, J. & Bonatto, D. (2011). The developmental aging and origins of health and disease hypotheses explained by different protein networks, *Biogerontology* 12: 293–308. 10.1007/s10522-011-9325-8.

Freeman, L. C. (1977). A set of measures of centrality based on betweenness, *Sociometry* 40(1): 35–41.

Gilbert, D. (n.d.). http://www.jfree.org/jfreechart/.

Hu, Z., Mellor, J., Wu, J., Yamada, T., Holloway, D. & DeLisi, C. (2005). VisANT: data-integrating visual framework for biological networks and modules, *Nucl. Acids Res.* 33(suppl_2): W352–357.

Jeong, H., Mason, S. P., Barabasi, A. L. & Oltvai, Z. N. (2001). Lethality and centrality in protein networks, *Nature* 411(6833): 41–42.

Jeong, H., Tombor, B., Albert, R., Oltvai, Z. N. & Barabasi, A. L. (2000). The large-scale organization of metabolic networks, *Nature* 407(6804): 651–654.

Joy, M. P., Brock, A., Ingber, D. E. & Huang, S. (2005). High-betweenness proteins in the yeast protein interaction network., *J Biomed Biotechnol* 2005(2): 96–103.

Junker, B., Koschutzki, D. & Schreiber, F. (2006). Exploration of biological network centralities with centibin, *BMC Bioinformatics* 7(1): 219+.

Kinexus (n.d.). http://www.kinexus.ca.

Koschützki, D., Lehmann, K. A., Peeters, L., Richter, S., Podehl, D. T. & Zlotowski, O. (2005). Centrality indices, *in* U. Brandes & T. Erlebach (eds), *Network Analysis: Methodological Foundations*, Springer, pp. 16–61.

Ladha, J., Donakonda, S., Agrawal, S., Thota, B., Srividya, M. R., Sridevi, S., Arivazhagan, A., Thennarasu, K., Balasubramaniam, A., Chandramouli, B. A., Hegde, A. S., Kondaiah, P., Somasundaram, K., Santosh, V. & Rao, S. M. R. (2010). Glioblastoma-specific protein interaction network identifies pp1a and csk21 as connecting molecules between cell cycle-associated genes., *Cancer Res* 70(16): 6437–47.

Lepp, Z., Huang, C. & Okada, T. (2009). Finding Key Members in Compound Libraries by Analyzing Networks of Molecules Assembled by Structural Similarity, *Journal of Chemical Information and Modeling* 0(0): 091030094710018+.

Milo, R., Shen-Orr, S., Itzkovitz, S., Kashtan, N., Chklovskii, D. & Alon, U. (2002). Network motifs: Simple building blocks of complex networks, *Science* 298(5594): 824–827.

Newman, M. E. J. (2006). Modularity and community structure in networks, *Proceedings of the National Academy of Sciences* 103(23): 8577–8582.

Scardoni, G. & Laudanna, C. (2011). Interference: a tool for virtual experimental network topological analysis.
    URL: *http://www.cbmc.it/ scardonig/interference/Interference.php*

Scardoni, G., Petterlini, M. & Laudanna, C. (2009). Analyzing biological network parameters with CentiScaPe, *Bioinformatics* 25(21): 2857–2859.

Schokker, D., de Koning, D.-J., Rebel, J. M. J. & Smits, M. A. (2011). Shift in chicken intestinal gene association networks after infection with salmonella., *Comp Biochem Physiol Part D Genomics Proteomics* .

Sengupta, U., Ukil, S., Dimitrova, N. & Agrawal, S. (2009a). Expression-based network biology identifies alteration in key regulatory pathways of type 2 diabetes and associated risk/complications., *PloS one* 4(12): e8100+.

Sengupta, U., Ukil, S., Dimitrova, N. & Agrawal, S. (2009b). Expression-based network biology identifies alteration in key regulatory pathways of type 2 diabetes and associated risk/complications., *PloS one* 4(12): e8100+.

Sengupta, U., Ukil, S., Dimitrova, N. & Agrawal, S. (2009c). Identification of altered regulatory pathways in diabetes type ii and complications through expression networks, *Genomic Signal Processing and Statistics, 2009. GENSIPS 2009. IEEE International Workshop on*, pp. 1 –4.

Shannon, P., Markiel, A., Ozier, O., Baliga, N. S., Wang, J. T., Ramage, D., Amin, N., Schwikowski, B. & Ideker, T. (2003). Cytoscape: a software environment for integrated models of biomolecular interaction networks., *Genome research* 13(11): 2498–2504.

Shen-Orr, S., Milo, R., Mangan, S. & Alon, U. (2002). Network motifs in the transcriptional regulation network of escherichia coli, *Nature Genetics* 31.

Strogatz, S. H. (2001). Exploring complex networks, *Nature* 410(6825): 268–276.

Venkatachalam, G., Kumar, A. P., Sakharkar, K. R., Thangavel, S., Clement, M.-V. & Sakharkar, M. K. (2011). Pparγ; disease gene network and identification of therapeutic targets for prostate cancer., *J Drug Target* .
    URL: *http://www.biomedsearch.com/nih/PPAR-disease-gene-network-identification/21780947.htr*

Wagner, A. & Fell, D. A. (2001). The small world inside large metabolic networks., *Proceedings. Biological sciences / The Royal Society* 268(1478): 1803–1810.

Watts, D. J. (1999). *Small worlds: the dynamics of networks between order and randomness*, Princeton University Press, Princeton, NJ, USA.

Watts, D. J. & Strogatz, S. H. (1998). Collective dynamics of 'small-world' networks, *Nature* 393(6684): 440–442.

Webster, Y. W., Dow, E. R., Koehler, J., Gudivada, R. C. & Palakal, M. J. (2011). Leveraging Health Social Networking Communities in Translational Research., *Journal of biomedical informatics* .
    URL: *http://dx.doi.org/10.1016/j.jbi.2011.01.010*

Wuchty, S. & Stadler, P. F. (2003). Centers of complex networks., *J Theor Biol* 223(1): 45–53.

Yamada, T. & Bork, P. (2009). Evolution of biomolecular networks: lessons from metabolic and protein interactions., *Nature reviews. Molecular cell biology* 10(11): 791–803.

6

Camera Motion Estimation Based on Edge Structure Analysis

Andrey Vavilin and Kang-Hyun Jo
University of Ulsan,
Korea

1. Introduction

The estimation of camera motion is important for several video analysis tasks such as indexing and retrieval purposes, motion compensation and for scientific film analysis. From an aesthetical point of view, camera motion is often used as an expressive element in film production. Motion content can be used as a powerful cue for structuring video data, similarity-based video retrieval, and video abstraction

Motion estimation and motion pattern classification problem have been extensively investigated by the scientific community for semantic characterization and discrimination of video streams. Moving object trajectories have been used for video retrieval [1–3]. Camera motion pattern characterization has been efficiently applied to video indexing and retrieval [4–7]. However, the main limitation of the latter methods is that they deal only with the characterization of the detected camera motion patterns, without explicit measurement of the camera motion parameters. As a result, the acquired information is of limited interest, since it can be used primarily for video indexing and retrieval.

There are different types of camera motion: rotation around one of the three axes and translation along the x and y-axis. Furthermore, zoom in and out can be considered as equivalent to translation along the z-axis. Existing methods can be classified as optical flow methods and feature correspondences based approaches. Let us also mention recursive techniques based on extended Kalman filters [8] which track camera motion and estimate the structure of the scene. In the case of an uncalibrated camera, interesting approaches are described in [9, 10]. The use of optical flow avoids the choice of "good features". In [11] differential approaches of the epipolar constraint are described. In [12], the optical flow computed between two adjacent images in a video sequence is linearly decomposed on a database of optical flow models. The authors of [13] propose a comparison of algorithms which only use optical flow for estimating camera.

In this work we present an approach for recovering graph-based structures from images. This structure is then used for estimating 3D camera motion. Our approach is based on detecting straight line segments. After several prefiltering operations such as bilateral filtering and Hough transform, preceding by the edge detection is applied. Then a result of transformation analyses in order to detect local maximums. Reverse transform of these maximums gives us a several straight lines presented in the image. We use an intersection of these analytical lines with detected edges in order to find straight line segments. On this step we also detect intersections between line segments. For each intersection point we compute rank based on number of connected points. After transforming image into graph we search for similar structural elements in the graph of previous frame. This process is based on searching subgraphs consists of vertices with similar ranks. After finding correspondence points camera motion is estimated as a combination of translation and rotation.

2. Camera motion model

In this work we considered eight-parameter perspective model defined as follows:

$$x_i^2 = \frac{a_0 + a_2 x_i^1 + a_3 y_i^1}{a_6 x_i^1 + a_7 y_i^1 + 1}$$
$$y_i^2 = \frac{a_1 + a_4 x_i^1 + a_5 y_i^1}{a_6 x_i^1 + a_7 y_i^1 + 1}$$

$$(1)$$

where $\left(x_1^1, y_1^1\right)$ and $\left(x_i^2, y_i^2\right)$ are the coordinates of the same point in the consequent frames at t_1 and t_2 respectively and $\left(a_0, ..., a_7\right)$ are the motion parameters. Various motion models can be derived from this mode. For example, in case of $a_6 = a_7 = 0$ it is reduced to affine model, and setting $a_2 = a_5$, $a_3 = -a_4$ and $a_6 = a_7 = 0$ will give us a translation-zoom-rotation model.

In [14] any vector field is approximated by a linear combination of a divergent field, a rotation field and two hyperbolic fields. The relationship between motion model parameters and symbol-level interpretation is established as:

$$Pan = a_0$$
$$Tilt = a_1$$
$$Zoom = \frac{1}{2}\left(a_2 + a_5\right)$$
$$Rotation = \frac{1}{2}\left(a_3 - a_4\right)$$

$$(4)$$

Error in esti mation parameters is defined as:

$$\varepsilon(a) = \sum_{i=1}^{N}\left\|p_i^2 - f\left(p_i^1, a\right)\right\|$$

$$(2)$$

where N is the number of corresponding points, $p_i^1 = \left(x_i^1, y_i^1\right)$ and $p_i^2 = \left(x_i^2, y_i^2\right)$ are the corresponding points in first and second frames, $a = \left(a_0, ..., a_7\right)$ is a transformation parameters vector and $f()$ is a transformation function defined by (1).

Using this model definition, problem of camera motion estimation could be formalized as the error minimization problem:

$$M = \arg\min_{a} \varepsilon(a) \tag{3}$$

Where M defines estimated camera motion parameters.

It is well known that, by taking some particular point p as the origin of the coordinate system with coordinates z, any infinitely differentiable function $f(x)$ could be approximated using Taylor series:

$$f(z) = f(p) + \sum_i \frac{\partial f}{\partial z_i} z_i + \frac{1}{2} + \sum_i \frac{\partial^2 f}{\partial z_i \partial z_j} z_i z_j + \ldots \approx c - b \cdot z + \frac{1}{2} z \cdot A \cdot z \tag{4}$$

where

$$c \equiv f(p) \qquad b \equiv -\nabla f \,|\, p \qquad [A]_{ij} = \frac{\partial^2 f}{\partial z_i \partial z_j} \,|\, p \tag{5}$$

The matrix A which consists of a second partial derivatives of the function is also called Hessian matrix of the function at p [15]

In approximation of (4) the gradient of f is easily calculated as

$$\nabla f(z) = Az - b \tag{6}$$

In Newton's method gradient is set to 0 to determine the next iteration point.

The gradient of error function ε with respect to parameters a has components

$$\frac{\partial \varepsilon}{\partial a_k} = -2 \sum_{i=1}^{N} \frac{\left[p_i^2 - f(p_i^1, a) \right] \partial f(p_i^1, a)}{\partial a_k}, \qquad k = 0, 1, \ldots, 7 \tag{7}$$

Taking second order partial derivatives gives

$$\frac{\partial^2 \varepsilon^2}{\partial a_k \partial a_j} = 2 \sum_{i=1}^{N} \left[\frac{\partial f(p_i^1, a)}{\partial a_k} \frac{\partial f(p_i^1, a)}{\partial a_j} - \left[p_i^2 - f(p_i^1, a) \right] \frac{\partial^2 f(p_i^1, a)}{\partial a_k \partial a_j} \right] \tag{8}$$

It is conventional to remove the factors of 2 by defining

$$\alpha_{kj} = \frac{1}{2} \frac{\partial^2 \varepsilon^2}{\partial a_k \partial a_j} \tag{9}$$

$$\beta_k = \frac{1}{2} \frac{\partial \varepsilon}{\partial a_k} \tag{10}$$

Making [a]=1/2A in equation (6), in terms of which that equation can be rewritten as the set of linear equations

$$\sum_{i=1}^{N} \alpha_{ki}\partial a_i = \beta_k \qquad (11)$$

SVD is used to compute transformation parameters form overdetermined set of linear equations (11).

In the proposed work initial translation was estimated prior to pan-tilt-zoom estimation based on center of gravity of corresponding feature points in consistent frames. To remove outliers voting idea was used. After finding correspondences between frames each pair of matching points "votes" for its offset. Then points with small number of offset votes are discarded.

3. Image to graph conversion

However, the most challenging part in camera motion estimation is finding correspondences between two frames. In proposed work this process is based on searching similar edge structures in the frames. Input image is represented as a graph based on edges and their intersections. Intersection points are ranked according to the number of connections. Algorithm for transforming image into graph is presented in Fig.1.

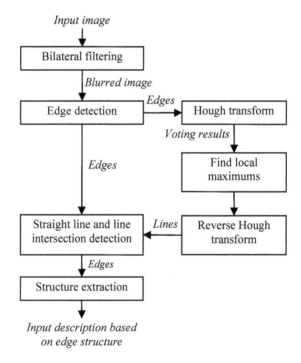

Fig. 1. Transforming image into graph

The proposed method is based on matching edge structures inside the images. Robustness of the algorithm depends on the quality of detected edges. Thus, we try to discard small edges and edges with small magnitude before recovering structures. One of the most effective way to remove such edges is a bilateral filtering [18]. We used bilateral filtering to blur image while preserving strong edges (Fig.2(b)). This allows us to reduce number of small edges. With a Gaussian function $g(x,\sigma) = \exp\left(-x^2 / \sigma^2\right)$, bilateral filter of input image I at pixel p is defined as

$$bf(I)_p = \frac{1}{k}\sum_{q\in I}\left[G\left(\|p-q\|,\sigma_s\right)\cdot G\left(\left|I(p)-I(q)\right|,\sigma_I\right)\cdot I(q)\right] \tag{5}$$

where σ_s controls the influence of spatial neighbourhood, σ_I the influence of the intensity difference and k is a normalizing coefficient defined as

$$bf(I)_p = \frac{1}{k}\sum_{q\in I}\left[G\left(\|p-q\|,\sigma_s\right)\cdot G\left(\left|I(p)-I(q)\right|,\sigma_I\right)\right] \tag{6}$$

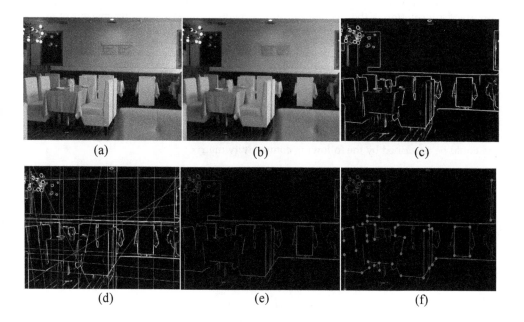

(a)   (b)   (c)

(d)   (e)   (f)

Fig. 2. Example of image to graph transformation. Refer text for details.

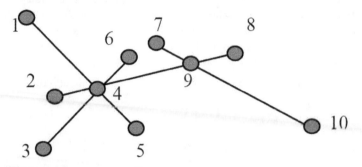

Fig. 3. Example of graph structure.

Next step is edge detection (Fig.2(c)). In this work Sobel edge detector was used. Edges with magnitude less than 50 were filtered out during this step. Remained edges were mapped into ρ,φ space using Hough transform. As a result 2-dimensional voting array was generated. Local maximums of this array correspond to strongest edges in the image. Using reverse Hough transform for these maximums we obtain set of lines (Fig.2(d)). Intersection of these lines gives us straight line segments (Fig.2(e)) and their intersections (Fig.2(f)). The purpose of these steps is to obtained graph-like structure. In further processing we use information about edge intersections (vertices) and their connections (edges). The computational time of the proposed algorithm might be reduced by removing Hough transform. It is not very important to detect exactly straight lines between vertices. The important information is their connectivity. However, using straight lines instead of edges allows more robust detection of edge intersection.

Information about edges could be represented in the form of a graph $<E,V>$ where $v=\{v_1,...,v_m\}$ is a set of vertices obtained from intersections and $E=\{e_1,...,e_n\}$ is a set of edges. Each vertex is described by its coordinates and rank which shows how many connections this vertex has. Edges of graph show connections between vertices. To compute rank of the vertex we used the following idea. Let's consider the structure presented in the Fig.3 as a graph, obtained after edge detection. It has nine edges $e_1...e_9$, ten vertices $v_1...v_{10}$ and could be described by the following connectivity matrix :

$$C = \begin{bmatrix} 0 & 0 & 0 & 1 & 0 & 0 & 0 & 0 & 0 & 0 \\ 0 & 0 & 0 & 1 & 0 & 0 & 0 & 0 & 0 & 0 \\ 0 & 0 & 0 & 1 & 0 & 0 & 0 & 0 & 0 & 0 \\ 1 & 1 & 1 & 0 & 1 & 1 & 0 & 0 & 1 & 0 \\ 0 & 0 & 0 & 1 & 0 & 0 & 0 & 0 & 0 & 0 \\ 0 & 0 & 0 & 1 & 0 & 0 & 0 & 0 & 0 & 0 \\ 0 & 0 & 0 & 0 & 0 & 0 & 0 & 0 & 1 & 0 \\ 0 & 0 & 0 & 0 & 0 & 0 & 0 & 0 & 1 & 0 \\ 0 & 0 & 0 & 1 & 0 & 0 & 1 & 1 & 0 & 1 \\ 0 & 0 & 0 & 0 & 0 & 0 & 0 & 0 & 1 & 0 \end{bmatrix}$$

Let's define j-th order rank of the i-th vertex as a number of vertices which could be reached from i-th vertex by j steps. First order rank of the i-th vertex shows how many vertices are directly connected to it. Thus, the v_1 will have first order rank $r_1^1 = 1$ and the v_4 will have $r_4^1 = 6$. Second order rank shows how many vertices could be reached from the i-th vertex in two steps. Thus, $r_1^2 = 6$ and $r_4^2 = 9$. Finally, n-th order rank shows how many vertices could be reached from i-th vertex in n steps. This ranking could be used to evaluate complexity and size of the substructures. Rank table for the graph presented in Fig.3 is shown in Table 1. Let's define the highest order of the rank $rMax$ as the minimal order which satisfies the following condition:

$$rMax = j \mid r_i^j = n \forall i \geq rMax \tag{7}$$

where n is the number of edges in graph.

The highest order for this graph is 3. Using higher order rank is useless for that graph as long as it would have same values. However, in common case, using higher order ranks allows to make matching process more effective by detecting more complex structures inside the image.

	Vertices									
	v_1	v_2	v_3	v_4	v_5	v_6	v_7	v_8	v_9	v_{10}
r_i^1	1	1	1	6	1	1	1	1	4	1
r_i^2	6	6	6	9	6	6	4	4	9	4
r_i^3	9	9	9	9	9	9	9	9	9	9

Table 1. Vertex ranks

These ranks have the following properties:

- Maximum rank could not be higher than number of edges.
- Maximum rank is equal to the number of edges if and only if the graph does not contain loops.
- Highest order of the rank shows diameter of the graph: maximum number of steps required to reach any vertex from any other.

After transforming two frames into graphs problem of finding correspondences between them could be reformulated as a problem of finding sub-graphs with similar structure. We start our search with finding correspondences for vertices with high rank. During this process spatial information could also be considered. Thus we try to find corresponding vertex which would have the same rank and will be located close to the reference vertex. In the future work we this step may be improved by searching correspondences for groups of vertices instead of matching them one by one. As the result of this step we obtain two set of points P and Q which are used to estimate camera motion as it was described.

Matching idea is based on searching similar substructures in the image graphs. Typically, images contain many simple substructures. These substructures have a small maximum rank order *rMax*. Furthermore, most of the vertices have small rank of first and second order, while very few of them are ranked higher than 6. Thus we start searching of correspondences from the structures with highest ranks. In the graph presented on Fig.3 such vertices are v_4 and v_9. They have ranks 6 and 4 respectively. As long as they are connected we will try to search for two connected vertices with ranks 4 and 6 in the next frame.

To simplify matching process we try to find substructures with more complex structure. This problem is solved by computing rank of vertices of higher orders.

Some edges could be lost in image sequence due to noise, camera movement and movement of objects in scene. It may cause differences in the structure of image graphs. To solve this problem the following ideas were used:

1. We tried to match structures located closer to the center of frame first in order to avoid loosing edges due to camera motion.
2. We consider that vertex i in one frame may correspond to vertex j in other frame even if their ranks are not exactly same. Thus, for each vertex in first frame we select several candidates which ranks are differs less than 30%. After candidates are selected for all vertices we try to choose best corresponding pairs using spatial information.

To decrease number of incorrect matches and to increase the preciseness of matching algorithm matched points were additionally compared using Cross-Hexagon Search (CHS) algorithm [19]. Coordinates of matched graph vertices were used a block centers and offset between matched points in consistent frames was used as initial offset. As long as CHS is used for verification of matched feature points, it can be changed to any block-based feature matching algorithm such as Diamond Search [20] or Three-Step Search [21].

4. Experimental results

All experiments were done on Intel Core 2 Duo with 2Gb memory under Borland Builder environment. Program was not optimized for the maximum performance, thus, computational time, shown in Table 2, could be decreased.

Stage	Average time (sec)
Bilateral filtering	0,31
Edge detection and Hough transform	0,2
Representing image as a graph	0,03
Ranking	Less than 0,01
Graph matching	0,17
Total	0,72

Table 2. Average computational time for 200 vertices.

To evaluate matching quality several types of experiments were done. First group of tests considered planar camera motion (translation and rotation around camera optical axis). Image sequences with camera smoothly by shifted by 20 cm in different directions and rotated by 20 degrees around its optical axis were made (see Fig.4 for example). To evaluate quality of camera motion estimation average and maximum absolute errors were computed. Results for all groups of tests are shown in Table 3.

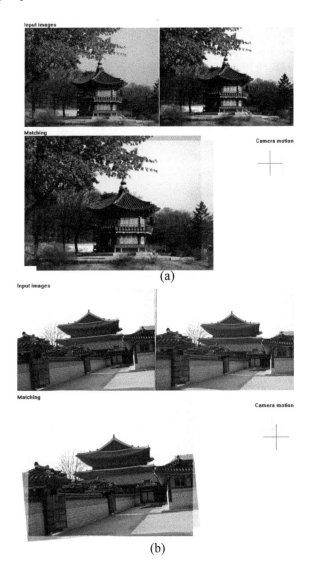

Fig. 4. Planar camera motion estimation for translation (a) and translation with rotation (b).

Second group of tests was used to evaluate algorithm performance for full 3D camera motion with known real camera trajectory. Three kinds of scenes were used: static scenes, scenes with moving objects and scenes with high amount of natural objects (trees, grass etc). Example of camera trajectory estimation is shown at Fig.5~6.

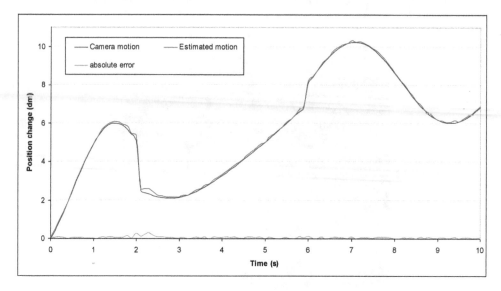

Fig. 5. Estimated trajectory for translation and zoom.

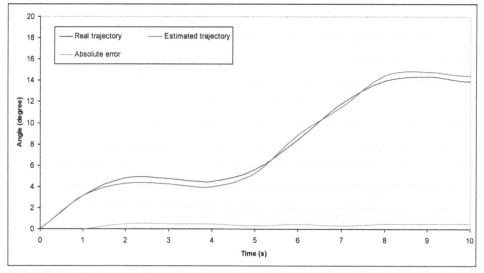

Fig. 6. Estimated trajectory for rotation.

Additionally, these tests were used to evaluate performance of algorithm with and without CHS correction. (Fig.7,8). It is easy to see, that using additional verification step for correspondent points could decrease estimation error, with relatively small increasing of computational time (about 0,07 sec).

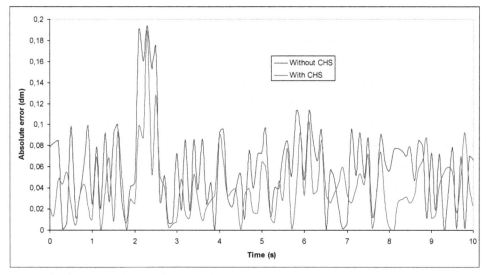

Fig. 7. Absolute error for translation and zoom.

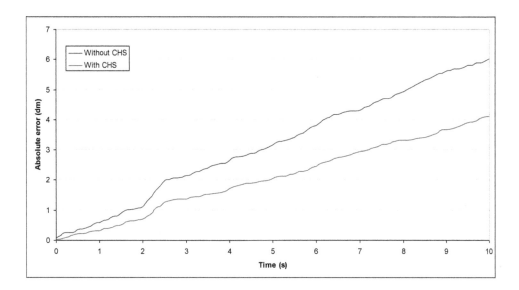

Fig. 8. Accumulated absolute error for translation and zoom.

In last group of tests image sequence with predefined camera motion trajectory was used to evaluate error depending on number of graph vertices used for matching result is shown in Fig.9.

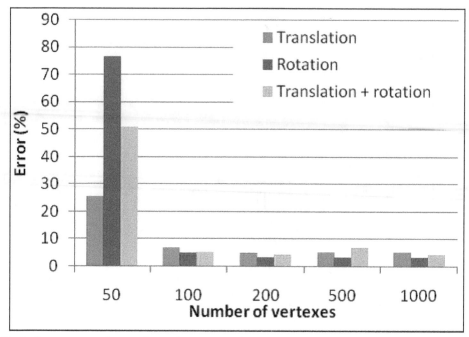

Fig. 9. Computational error depending on number of vertices.

	Planar motion	3D motion			3D motion with CHS		
		Static scene	Moving objects	Natural scene	Static scene	Moving objects	Natural scene
Average absolute error	0,59	0,44	0,59	4,61	0,33	0,37	4,3
Maximum absolute error	0,6	3,7	3,93	9,02	2,21	2,47	8.98

Table 3. Computational error (cm)

Table 3 shows that the proposed method can effectively estimate camera motion for scene with moving objects. However, its weak point is natural scenes with small number of geometrical objects.

5. Conclusion

In this paper we introduce an algorithm for converting images into graphs based on edges structure and graph matching algorithm based on vertex ranking. Proposed method could be used in different applications such as camera motion estimation, stereo matching, motion compensation, background model generation and many others. The proposed method provides efficient image-to-graph mapping for urban scenes. In case of natural scenes additional prefiltering may be required for removing unimportant information from edge image. In this paper we also presented simple algorithm for camera motion estimation based on parametric motion model. In future works we would like to improve the performance of the proposed algorithm to work in a real-time applications. Furthermore, algorithm could be improved for natural scenes by using different type of features.

6. Acknowledgments

This research was supported by the MKE(The Ministry of Knowledge Economy), Korea, under the Human Resources Development Program for Convergence Robot Specialists support program supervised by the NIPA(National IT Industry Promotion Agency) (NIPA-2010-C7000-1001-0007) and post-BK21 at University of Ulsan.

7. References

[1] W. Lie and W. Hsiao, "Content-based video retrieval based on object motion trajectory," in Proc. IEEEWorkshop on Multimedia Signal Processing, Dec. 2002, pp. 237–240.

[2] W. Hu, D. Xie, Z. Fu, W. Zeng, and S. Maybank, "Semantic-based surveillance video retrieval," IEEE Trans. Image Process., vol. 16, no. 4, pp. 1168–1181, Apr. 2007.

[3] Y. Jianfeng and L. Zhanhuai, "Modeling of moving objects and querying videos by trajectories," in Proc. 10th Int. Multimedia Modelling Conf., Washington, DC, 2004, pp. 373-380.

[4] Y.-P. Tan, D. D. Saur, S. R. Kulkarni, and P. J. Ramadge, "Rapid estimation of camera motion from compressed video with application to video annotation," IEEE Trans. Circuits Syst. Video Technol., vol. 10, no. 1, pp. 133–145, Jan. 2000.

[5] L.-Y. Duan, J. S. Jin, Q. Tian, and C.-S. Xu, "Nonparametric motion characterization for robust classification of camera motion patterns," IEEE Trans. Multimedia, vol. 8, no. 2, pp. 323–340, Apr. 2006.

[6] X. Zhu, A. K. Elmagarmid, X. Xue, L. Wu, and A. C. Catlin, "InsightVideo: Toward hierarchical video content organization for efficient browsing, summarization and retrieval," IEEE Trans. Multimedia, vol. 7, no. 4, pp. 648–666, Aug. 2005.

[7] T. Lertrusdachakul, T. Aoki, and H. Yasuda, "Camera motion characterization through image feature analysis," in Proc. ICCIMA'05, Aug. 16–18, 2005, pp. 186–190.

[8] A. Yao and A. Calway, "Robust estimation of 3-d camera motion for uncalibrated augmented reality", Technical Report CSTR-02-001, Dept of Computer Science, University of Bristol, 2002.

[9] O. Faugeras, Q.T. Luong, and T. Papadopoulo. "The Geometry of Multiple Images". MIT Press, 2000.

[10] M. Pollefeys, M Vergauwen, K Cornelis, J. Tops, F. Verbiest, and L. Van Gool "Structure and motion from image sequences" Proceedings of Conference on Optical 3-D Measurement Techniques V, pp. 251-258, 2001.

[11] Y. Ma, J. Koseck_a, and S. Sastry. "Linear differential algorithm for motion recovery: A geometric approach", IJCV, vol.36(1), pp.71-89, 2000.

[12] S.C. Park, H.S. Lee, and S.W. Lee, "Qualitative estimation of camera motion parameters from the linear composition of optical flow", PR(37), vol. 4, pp.767-779, 2004.

[13] Y. Tian, C. Tomasi, and D.J. Heeger, "Comparison of approaches to egomotion computation" IEEE Computer Society, Conference on Computer Vision and Pattern Recognition, pp.315-320, 1996.

[14] E. Francois and P.Bouthemy, "Derivation of qualitative information in motion analysis," Image Vi-sion Computing, vol.8, no.4, pp. 279-287, Nov.1990.

[15] W.H. Press, B.P. Flannery, S. A. Teukolsky, and W.T. Vetterling, Numerical Recipes in C: The Art of Scientific Computing. Cambridge, U.K.: Cambridge Univ. Press, 1988, pp.59-70, pp.656-706.

[16] R. Grompone von Gioi, J. Jakubowicz, J-M.Morel and G.Randall, "On Straight Line Segment Detection", Journal of Mathematical Imaging and Vision, vol.32(3), pp.313-347, November 2008.

[17] C. Beumier, "Straight-Line Detection Using Moment of Inertia", IEEE International Conference on Industrial Technology, pp.1753-1756, 15-17 Dec. 2006.

[18] C. Tomasi and R. Manduchi, "Bilateral Filtering for Gray and Color Images", Proceedings of the 1998 IEEE International Conference on Computer Vision, Bombay, India, 1998.

[19] S S. Zhu, J. Tian, X. Shen, K. Belloulata,"A new cross-diamond search algorithm for fast block motion estimation", the IEEE Int. Conf. Image Processing, *ICIP2009*, Cairo, Egypt, 07-11 Nov. 2009, Vol. I, pp. 1581-1584.

[20] Belloulata, K., Shiping Zhu, Jun Tian, Xiaodong Shen, "A novel cross-hexagon search algorithm for fast block motion estimation", Systems, Signal Processing and their Applications (WOSSPA), 2011 7th International Workshop on, On page(s): 1 - 4, Volume: Issue: , 9-11 May 2011

[21] Li Ren-xiang, Zeng Bing, Liou M. L, "A new three-step search algorithm for block motion estimation," IEEE Transactions on Circuits and Systems for Video Technology, vol. 4, pp. 438-442, April 1994.

7

Graph Theory for Survivability Design in Communication Networks

Daryoush Habibi and Quoc Viet Phung
Edith Cowan University
Australia

1. Introduction

Design of survivable communication networks has been a complex task. Without establishing network survivability, there can be severe consequences when a physical link fails. Network failures which may be caused by dig-ups, vehicle crashes, human errors, system malfunctions, fire, rodents, sabotage, natural disasters (e.g. floods, earthquakes, lightning storms), and some other factors, have occurred quite frequently and sometimes with unpredictable consequences. To tackle these, survivability measures in a communication network can be implemented at the service layer, the logical layer, the system layer, and the physical layer.

The physical layer is the base resource infrastructure of the network, and to be able to protect it, we need to ensure that the physical topology of the network has sufficient link and node diversity. Without this, protection at higher layers will not be feasible. With the implementation of Dense Wavelength Division Multiplexing (DWDM) in the optical backbone of metropolitan and long-haul networks, greater flexibility is achieved in providing alternate routes for light-path connections. However, the survivability problem at the physical layer remains the same. In fact, it becomes even more critical, because each link of a backbone network carries huge amounts of traffic and the failure of an optical component, such as a fiber cut or a node failure, may cause a very serious problem in terms of loss of data and profit.

The physical topology of a network is considered to be survivable if it can cope with failure scenarios occurring at network components. In other words, the physical topology must remain connected under the failure scenario. For example, to cope with single link failures in the network, the physical topology must be at least 2-connected, meaning that there is at least 2 link-disjoint paths between any two nodes in the network. Generally, to protect against the failure of any set of k links in a network, the physical topology of that network must be $(k + 1)$-connected. Menger's theorem (Menger, 1927) gives the necessary and sufficient condition for survivability of networks at the physical layer, using the connectivity between network's cut-sets. However, the computational complexity of this model grows exponentially with the size of the network, since a network with $|V|$ nodes would yield $2^{|V|} - 2$ cut-sets. Therefore, the cut-set technique cannot efficiently deal with even moderate size networks of say 40 nodes, and larger networks are out of computational reach of this technique. Testing for survivability of large networks can be done using a technique called bi-connected components of a graph introduced by W. D. Grover (Grover, 2004). This technique can determine vulnerable links and nodes of the network. However, verifying network survivability is just the first step in network planning, after which we need to apply appropriate protection routing schemes using such techniques as Shared Backup Path Protection (SBPP), Pre-configured Protection Cycles (p-cycle), or ring protection. It is therefore very helpful if the algorithm used for determining the physical survivability of the network can also provide additional information which is of benefit to protection design.

2. Chapter outline

It is generally understood that research in transport networks has a close connection to graph theory. A graph can be used to present key aspects of a network, such as its topology and/or the associated capacity. Indeed, a network consists of many more elements such as the physical equipment (e.g. OADMs, OXCs, etc.), fibre cables, and so on. The objective of this chapter is to provide an overview of network connectivity in relation to network protection design. In addition, this chapter also aims to introduce and analyse the advantages and disadvantages of methods and algorithms for searching network connectivity as well as sets of disjoint and distinct paths for protection design.

Given these objectives, the chapter is organized as follows. Section 3 discusses the connectivity property of graph in relation to network protection design. Section 4 discusses two approaches to establishing the survivability of physical topology. In Section 5, we present the problem of diverse routing and graph algorithms. Finally, we close this chapter in Section 6 by summarising the key aspects of network connectivity in network protection design.

3. Graph connectivity in relation to network protection design

Practically, different traffic requirements over a network would require different connectivity between nodes. Some traffic demands may require no protection, or may only need to be carried when possible. In contrast, other traffic demands may ask for a full protection against either single-link failure, dual-link failures or other types of failure. Hence, the required connectivity between nodes would be different for these varied protection requirements. This section first presents the general requirement for network connectivity in which the network can at least be protected against any single failure scenario, e.g. the failure of any single link or node of the network. In this case, the physical topology of the network must be 2-connected.

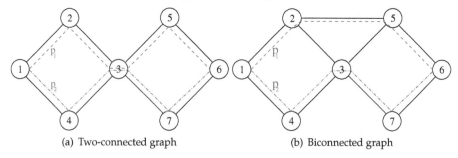

(a) Two-connected graph   (b) Biconnected graph

Fig. 1. Two connected graph versus biconnected graph

The protection mechanism is designed to maintain the continuity of services under network failures. The failure can be minor such as a channel failure, or be more significant such as failure of groups of links or nodes. As far as the connectivity of the physical topology is concerned, the failure may be in a single link, a single node, a group of links, a group of nodes. Protection design cannot cover all failure scenarios. Instead, it will consider the set of specific (or predetermined) failures. For example, protection design against single link failures needs to determine the recovery routes for services so that they can maintain their services under the failure of any single link in the network. To do that, the network connectivity must provide at least 2 *disjoint* paths between any source and destination nodes. The term "disjoint" here is with respect to the failure scenario, meaning that the "disjoint" paths must not suffer from the same failure. For example, for single link failures, the 2 paths must be link-disjoint. Similarly, for single node failures, the 2 paths must be node-disjoint.

Fig. 1 shows an example of a pair of link-disjoint paths (Fig. 1(a)) and node-disjoint paths (Fig. 1(b)). It can be easily seen that link-disjoint paths may share nodes along their paths, e.g. node 3 on paths p_1 and p_2 in Fig. 1(a). Hence, link-disjoint paths may not be node-disjoint. On the other hand, node-disjoint paths are always link-disjoint.

A graph which provides at least 2 link-disjoint paths between any two nodes is 2-*connected*. With stronger connectivity, a *biconnected* graph is able to provide at least 2 node-disjoint paths between any two nodes.

Generally, a network must provide at least K link-disjoint paths between any node-pairs to be able to protect against simultaneous failure of $K - 1$ links. The graph of such networks is said to be K-*connected*. The rest of this chapter will mainly discuss network connectivity which supports single link and single node failures, known as 2-connected and biconnected.

4. Establishing physical survivability

This section presents the procedures for establishing network survivability against single-link failures which is one of the most common failure scenarios occurring in practical networks. As discussed, the physical topology of a network can only be survivable under single-link failures if and only if it is 2-connected. Manual verification for survivability is only suitable with small networks where designers can perform the verification in just few seconds or few minutes. However, manual verification may take hours or even days for large scale networks. Furthermore, it is prone to human errors. Therefore, automatic survivability verification is important in both theory and practice. The concept of survivable networks is more complex than the concept of connectivity in graph theory. In addition, efficient automation algorithms based on graph theory can help designers to reduce the computational time and avoid human errors. This section presents techniques for evaluating the physical survivability of networks. Firstly, we outline and analyse the strengths and weaknesses of a popular method, namely the cut-set method. Then, we introduce two comparable techniques that can deal with network sizes of many thousand nodes (Habibi et al., 2005). One technique is based on Depth-First Search (DFS) and the other uses properties of 2-connected graphs.

4.1 Survivability via cut-sets

A network is survivable if the size of every cut-set of the network is equal to or larger than 2. At a glance, this definition leads to a view that the network has nodal-degree of two, meaning that every node in the network is connected to at least two other nodes. Since every node is connected to at least two other nodes in the network, on the surface this property seems to be able to offer two disjoint paths between any two nodes in the network. In fact, this is a misconception. If a network is 2-connected then the nodal degree of all nodes in the network is equal to or larger than 2. The reverse does not hold, however, in that a network in which the nodal degree of all its nodes equal to or larger than 2 is not always 2-connected. The topology in Fig. 2 illustrates this concept. In that Figure we can see that path $(3 - 5 - 6)$ is a bridge that connects two subsets of network nodes $X = \{1, 2, 3, 4\}$ and $Y = \{6, 7, 8, 9\}$. As a result, all paths between nodes $x \in X$ and $y \in Y$ must share the same path $(3 - 5 - 6)$. Hence, although all nodes in this network have a nodal degree equal to or larger than 2, it is not a 2-connected network.

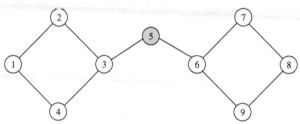

Fig. 2. An illustration of the failure of the nodal degree technique

Therefore, all algorithms for verification of network survivability based on node-degree of two may yield undesirable and inaccurate results and hence are not reliable. The cut-set assumption described below has been preferred for the accuracy of network survivability verification.

Let $G = (V, E)$ be a network topology. A cut in G is a partition of V into parts S and $\bar{S} = V \setminus S$. Each cut defines a set of edges consisting of those edges in E with one end-point in S and the other in $\bar{S}$. This edge set is referred as the cut-set $CS(S, V \setminus S)$ associated with the cut $\langle S, V \setminus S \rangle$. Let $|CS(S, V \setminus S)|$ be the size of the cut-set, being the number of links between S and $V \setminus S$. Thus, according to the cut-set assumption, **a network is 2-connected if** $|CS(S, V \setminus S)| \geq 2, \forall S \subset V$. If S is a subset of only a single node in the network, then the cut-set assumption is essentially the same as the node-degree assumption.

Since the cut-set assumption is related the number of links connected between two subsets of a cut, it can assure the network to offer **link-disjoint paths**, but **not node-disjoint paths**. In other words, a configuration of the network that satisfies the condition of the cut-set assumption can provide at least one link-disjoint path-pair between any distinct pair of source node and destination node.

The implementation of the cut-set assumption is not complex but its computational time for large scale networks is its biggest disadvantage. The number of cut-sets increases exponentially with the number of network nodes and is calculated as (Grover, 2004):

$$N_{cutset} = 2^{|V|} - 2$$

where N_{cutset} is the number of cut-sets in the network, and $|V|$ is the number of network nodes.

Table 1 shows the example of the number of the possible cut-sets, N_{cutset}, versus the number of network nodes $|V|$. The number of cut-sets doubles with an increase of one node in the network. For instance, N_{cutset} in a network of 20 nodes is over 1 million; and it is over 32 million with $|V| = 25$; which is $32 (= 2^5)$ times larger than $|V| = 20$; and the number of cut-sets in the networks of $|V| = 30$ nodes is up to 1 billion cut-sets. So, the cut-set technique becomes intractable even with moderate scale networks ($20 \leq |V| \leq 30$).

$|V|$	20	25	30
N_{cutset}	$1,048,574$	$33,554,430$	$1,073,741,822$

Table 1. The number of cut-sets versus the number of network nodes

In summary, the node-degree assumption is simple but not reliable for the verification of network survivability. Meanwhile, the cut-set assumption is only applicable for link-survivable networks, and it is intractable with large scale networks. The verification ability of these two assumptions for different connectivities of the physical topology are summarised in Table 2. The node-degree assumption cannot verify any type of physical topology that has potential to support network survivability (namely 2-connected and biconnected networks) whereas the cut-set assumption can verify the survivability of a network that is 2-connected but cannot identify exactly a 2-connected topology or verify a biconnected topology. Next, we propose an approach that can classify network topologies, and determine if they are unconnected, $(1-)$connected, 2-connected or biconnected.

	Network connectivity			
	Disconnected	(1-)Connected	2-connected	Biconnected
Node-degree assumption	Yes	Yes	No	No
Cut-set Assumption	Yes	Yes	Yes	No

Table 2. Performance of two common assumptions in terms of network connectivity

4.2 An approach to verifying physical topology for designing network survivability

Let $G(V, E)$ be the graph presenting the physical topology of a network. If the degree of any node in G is zero, the graph is disconnected. If the degree of any node in G is 1, the graph cannot be 2-connected or biconnected. For the sake of network survivability design, from here after, we assume that the degree of every node in the graph is at least 2. This condition allows every node to belong to a biconnected component or a bridge. Let G' and G'' be two biconnected components of the graph G. The relationship between G' and G'' determines the connectivity of the graph:

- If G' and G'' have at least 2 common nodes, then G is a 2-connected graph with no cut-node (i.e. node bridge) or cut-link (i.e. link bridge). In other words, G is a biconnected graph.
- If G' and G'' only have one common node, then G is a 2-connected graph with an *articulation node* which is the common node.
- If G' and G'' are separated by a cut-link (or a cut-path), then G is not a 2-connected graph, and the cut-link (or cut-path) cannot be protected.
- If G' and G'' have no common links or nodes, then G is a disconnected graph.

The key point for determining the connectivity of a graph is to find all biconnected components in the graph and check the relationship between these components. Biconnected components can be found using either Depth-First Search (DFS) or the properties of the 2-connected graph.

4.2.1 Finding biconnected components using Depth-First Search (DFS)

Algorithm 1 outlines the pseudo-code for verifying a 2-edge-connected graph. The principle is to find a bridge (edge) via articulation nodes - a cut at one of these nodes would disconnect the graph. A bridge is an edge or a path segment that connects two biconnected component neighbourhood. A graph is 2-edge-connected if it contains no bridge. The DFS algorithm for finding biconnected components was proposed by Robert Tarjan (Tarjan, 1971), and subsequently, revised for a particular study in survivable mesh networks in (Grover, 2004). Algorithm 1 is a modified version of the original DFS algorithm in (Grover, 2004; Tarjan, 1971) that allows for the verification of the 2-edge connectivity of a physical topology.

The underlying concept behind this search is to firstly "depth" explore unvisited nodes, called "depth search", via a walk through incident edges from the current node (v). At any depth search, the current visited node is pushed to a stack. The depth search continues until there are no unvisited nodes reachable from the current node. The search is "backtracking" by examining visited nodes in the stack and continuing depth search at these nodes. The algorithm is terminated when there are no nodes in the stack.

Algorithm 1 Depth-First Search for 2-connected graph

Require: A graph $G(\mathbf{V}, \mathbf{S})$ presented the physical topology of a given network.
Ensure: Return **true** if G is two connected, otherwise **false**
1: $count \leftarrow 1$;
2: Set current node $v \leftarrow v_0$; //start at node v_0
3: Set $dfs[v] = btk[v] \leftarrow count + +$;
4: **repeat**
5:     **if** (there is an unvisited incident edge e of v) **then**
6:         **if** ($\exists w \leftarrow v$ is unvisited) **then**
7:             $dfs[v] = btk[v] \leftarrow count + +$;
8:             Push $stack \leftarrow v$;
9:             Current node $v \leftarrow w$;
10:        **else**
11:            $btk[w] = min(btk[w], btk[v])$
12:        **end if**
13:    **else**
14:        Pop $w \leftarrow stack$; //parent of node
15:        **if** ($btk[v] \geq dfs[w]$) **then**
16:            Record w is an articulation node
17:        **else**
18:            $btk[w] = \min btk[w], btk[v]$;
19:        **end if**
20:    **end if**
21:    Current node $v \leftarrow w$;
22: **until** (there is no node in the stack)
23: **if** (there is no edge bridge between any two articulation nodes) **then**
24:     **return true**;
25: **else**
26:     **return false**;
27: **end if**

Two parameters are assigned to each visited node v, the $dfs[v]$ and the $btk[v]$, to discover articulation nodes. The $dfs[v]$ identifies the "depth first search" order of the node v when it is visited for the first time. The $btk[v]$ is determined using "back tracking" as follows:

- Initially, $btk[v]$ is assigned a value equivalent to the value of $dfs[v]$ when the node is first visited.
- $btk[v]$ is then updated as $btk[v] = \min(btk[v], dfs[w])$ when it has a neighbour w through an unvisited edge, and w is an ancestor of v.
- The value of $btk[w]$, where w is the parent of v, is also updated during the backtracking step as $btk[w] = \min(btk[w], btk[v])$.

The calculation of the value btk helps determine if a node v is a cut node (or articulation node) when $btk[v]$ is larger than or equal to $dfs[w]$, where w is the parent of v.

4.2.2 Finding biconnected components using property of 2-connected graphs

Alternatively, based on the properties of 2-connected graphs, biconnected components can be built from a simple and small cycle by adding the so-called H-paths to the cycle. Let H be a biconnected component on the graph G. A H-path is a non-trivial path on G that meets H exactly at its end nodes. The following proposition and proof are adopted from (Diestel, 2000).

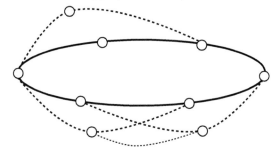

Fig. 3. The construction of 2-connected graphs

Proposition 4.1. *A graph is 2-connected if and only if it can be constructed from a cycle by successively adding H-paths to graph H already constructed.*

Proof. Clearly, every graph constructed as proposed is 2-connected. Conversely, let G be a 2-connected graph, then G contains a cycle, and hence a maximal subgraph H is constructable, as evident in Fig. 3. Any edge $x, y \in E(G) \setminus E(H)$ with $(x, y) \in H$ defines a H-path. Then, H is an induced sub-graph of G. If $H \neq G$, then by the connectedness of G, there is an edge vw with $v \in G - H$ and $w \in H$. As G is 2-connected, $G - w$ has a $v - H$ path P. Then wvP is a H-path in G, and $H \cup wvP$ is a constructable sub-graph of G. □

Based on this proposition, we can use the relationship between network's cycles or 2-connected graphs to verify the survivability of its physical topology.

An undirected graph is thus seen as the combination of all fundamental cycles. Using Algorithm 2, these fundamental cycles can be found from a spanning tree $T(V, E')$, $E' \subset E$ of a graph $G(V, E)$ (eg. the spanning tree highlighted by thick lines in Fig. 4).

Algorithm 2 Finding cycles

Require: A tree T and and edge e whose end-nodes is in T.
Ensure: A cycle P formed by T and e.
1: $(s, d) \leftarrow$ end-nodes of e;
2: $queue \leftarrow [node.s, node.P]; check \leftarrow 0$;
3: **while** $(check = 0 \& queue \neq \varnothing)$ **do**
4:      $[v] \leftarrow head(queue)$;
5:      $queue \leftarrow queue - head(queue)$;
6:      **if** $(v.s = d)$ **then**
7:          $check = 1; P \leftarrow v.P$;
8:      **else**
9:          **for** (all v_k is neighbour of v_s) **do**
10:             $node.s \leftarrow v_k; node.P \leftarrow P \cup v_k$;
11:             push $node$ into $queue$;
12:          **end for**
13:      **end if**
14: **end while**

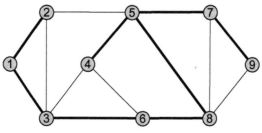

Fig. 4. Spanning tree on an arbitrary graph

However, it is not easy to find all of the fundamental cycles in the graph. For instance, in Fig. 4, the edges represented with thin lines are not part of the spanning tree (shown by thick lines). If any of these edges are added to the tree, it will form a unique cycle, but such cycle is not necessarily a fundamental cycle (eg. consider adding edge $2 - 5$). Any set of cycles found from the spanning tree can be used to verify the survivability of the topology from which it is generated. An algorithm for finding a set of cycles through spanning tree of a graph is represented in Algorithm 2. An efficient method for finding fundamental cycles of a graph, referred to as *Paton's* algorithm, is outlined in (Paton, 1969). Further discussion of this topic is outside the scope of this chapter.

The connectivity of the graph then can be determined using Algorithm 3. Note that the **set of remaining links** contains links in the graph but not in the tree.

5. Graph algorithms for diverse routing in protection design

Previously in this chapter, we have discussed the connectivity of the physical topology to support the problem of designing multiple quality of protections. However, the assurance of connectivity at the physical layer is only the first step to ensure traffic demands can be conveyed from the source to the destination. Once a physical layer is given, the main goal of the Logical Topology Design (LTD) is to determine routes for traffic flow in the most efficient

---

Algorithm 3 Construct biconnected components

Require: A spanning tree T of the graph G and the set of remain links.
Ensure: Set of biconnected components.
1: Firstly, a biconnected subgraph $G_i, i = 1$ is found by picking up a link e in the set of remain links and joining it with a unique path in the tree between end nodes of e.
2: a new biconnected subgraph $G_j, j = i + 1$ is constructed from a remain link and a unique path between end nodes of that link. If G_j have more than one node in common with a component $G_k, k = 1 \ldots i$, join $G_k \leftarrow G_k \cup G_j$. Otherwise, a new biconnected subgraph G_j is recorded, $i \leftarrow i + 1$.
3: Repeat step 2 until there is no remain links left.
4: The set of constructed biconnected components $G_1, G_2, \ldots, G_i$ is return.

---

way, that is to minimise the amount of physical resources utilised, particularly the amount of capacity utilisation. One approach is to enumerate the set of best candidates for each demand. In network protection design, these candidates can be distinct paths or disjoint paths. The disjoint paths are to ensure the success of the recovery process from network failures. On the other hand, the distinct paths provide the flexibility in selecting the most efficient working and backup routes. This section presents graph algorithms to support the diverse routing problem including finding K shortest (least cost) paths, finding two disjoint paths and finding K shortest disjoint path-pairs between any two nodes in a network.

5.1 Algorithm for finding K shortest paths between two nodes in the network

A set of all paths between two nodes can be used to select the best paths on which traffic flows between these nodes may be carried. All paths between two nodes in the graph can be found using either the Breath-First Search or the Depth-First Search. However, the number of all paths increases exponentially with the size of the network. Hence, finding all paths for routing is not a practical approach for real-time provisioning. In addition, the number of best selected paths is very small compared to all possible paths. Therefore, finding a small set of best eligible paths would be more practical and computationally efficient.

This section discusses and analyses the complexity of algorithms presented in the literature. We outline an efficient graph algorithm for finding K shortest paths between any two nodes in the network. This algorithm is an adaptation from (Martins et al., 1998) for directed graph to undirected graph.

Model selection used in survivable network design employs a set of potential candidates as inputs and selects optimal candidates as outputs. Input candidates differ depending on the protection techniques used. This can be either a set of paths for routing working flows, backup routes (in non-joint optimisation approaches), a set of disjoint path-pairs or a set of cycles in p-cycle design. This section introduces a basic algorithm used as a subroutine for finding input candidates for routing working flows and backup routes in non-joint optimisation approaches; the K shortest (minimum) paths algorithm.

K shortest paths is a classical graph problem that has been widely studied (Eppstein, 1998; Martins et al., 1998; Martins & Santos, 2000). Eppstein *et al.* (Eppstein, 1998) proposed a k shortest path algorithm between any two nodes in a digraph in time complexity of $O(|E| + |V| \log |E| + K)$, where $|V|$ is number of nodes, $|E|$ is the number of edges and K is number of paths required. Martins *et al.* (Martins et al., 1998) present two algorithms for

---

Algorithm 4 K shortest paths algorithm

Require: An undirected graph $G(V, E)$, a pair of source and destination nodes and the number of shortest paths required.

Ensure: A set of K - shortest paths over graph G from s to d.

1: To assure the possible repetition of the algorithm in a path between a pair of source-destination nodes (s, d), the given network is enlarged with a super source node S and super destination D, with zero cost edges $(s*, s)$ and $(d, d*)$. Following this, the shortest tree from source node s to other nodes in the network is obtained, and the first shortest path is marked as $p_1 = \{s_0(= s), s_1, \ldots, s_{r-1}, s_r(= d)\}$ from s to d.

2: Determine the first node s_h in p_1 such that s_h has more than a single incoming edges. If a node s_h', of which the incoming edges are the incoming edges of s_h except those coming from s_{h-1}, does not exist, then generate the node s_h', else determine the next node s_i in p_1 that has not alternate yet. The cost $d(s, s_h')$ of shortest path from s to s_h' is calculated as:

$$d(s, s_h') = \min_x \left(d(s, x) + d(x, s_h') \right)$$

where (x, s_h') are incoming edges of s_h'.

3: For each $s_j = \{s_i, \ldots, s_{r-1}\}$, generate s_j' following the same rules as s_h', but with one more incoming edge of (s_{j-1}', s_j'). Clearly, the shortest path from s to s_j' is the second shortest path from s to s_j. Therefore $p_2 = \{s_0, \ldots, s_i, \ldots, s_{r-1}', s_r(= d)\}$ is the second shortest path. Repeat step 2 to determine next shortest path $p_k (k = 2, 3, \ldots)$ until $k = K$.

---

the K shortest paths problem, one based on a label setting algorithm and another based on a label correcting algorithm. The results show that these algorithms perform better than the algorithm in (Eppstein, 1998). The K-shortest path algorithm employed in this study is adapted from (Martins & Santos, 2000), which is proposed over a directed graph approach. Building on this idea, this study has developed an algorithm (Algorithm 4) which is relevant for an undirected approach as well. This algorithm runs with the time complexity of $O(K \times |E|)$, where K is the number of shortest paths required and $|E|$ is the number of undirected edges in the graph.

5.2 Algorithm for finding two disjoint paths (disjoint path-pair) between nodes in the network

Finding a disjoint path-pair between two nodes in the network is a basic solution for protection design in which the primary and secondary paths must not suffer from the same failure. Basically, finding a pair of disjoint paths can be done in two steps. In the first step, the first path is determined from the original graph. Then, all edges contained in the first path are removed from of the graph. The second disjoint path is determined from the residual graph. Both paths are usually determined using any shortest path algorithm such as Dijkstra's algorithm (Dijkstra, 1959) or Bellman-Ford algorithm (Bellman, 1958). The two step approach is simple in both concept and implementation. There is, however, no guarantee that the total cost of the found disjoint path-pair is minimum. In addition, this approach may fail to find a solution in some cases even when a disjoint path-pair exists between the two nodes. An example of a trap topology is shown in Fig. 5. It is easy to see in Fig. 5(a) that there are two disjoint paths between nodes 1 and 4. However, by using the two-step approach, the first shortest path may be path p_1 ($1 \rightarrow 2 \rightarrow 3 \rightarrow 4$) as shown in Fig. 5(b). The second path is determined after all spans contained in the first path (edges $\{(1-2),(2-3),(3-4)\}$) are removed. It can easily be seen that the second path can not be found since the graph is disconnected between nodes 1 and 4.

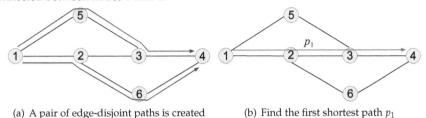

(a) A pair of edge-disjoint paths is created   (b) Find the first shortest path p_1

Fig. 5. An example of trap topology

To resolve these two drawbacks, a graph algorithm, known as the one step approach, for determining a shortest disjoint path-pair was first proposed by Surballe (Surballe, 1974) and modified by Bhandari (Bhandari, 1994; 1999) to adapt to the negative weight of edges. The algorithm, as its name implies, determines the first and the second disjoint path simultaneously.

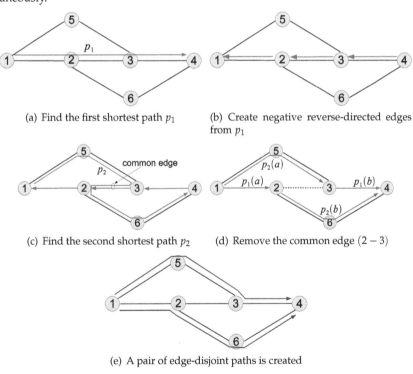

(a) Find the first shortest path p_1 (b) Create negative reverse-directed edges from p_1

(c) Find the second shortest path p_2 (d) Remove the common edge $(2-3)$

(e) A pair of edge-disjoint paths is created

Fig. 6. An illustration of the one-step approach to resolve trap topology

Let us assume that a shortest disjoint path-pair needs to be determined between the source node s and the destination node d. The one step algorithm is outlined as follows:

- Find a shortest (least cost) path between the source node s and the destination nodes d, and denote it as p_1 (Fig. 6(a)).
- Mark the direction of each edge traversed in p_1 from s toward d as positive.
- Remove all directed edges on the shortest path p_1 and replace them with reverse direction edges by multiplying -1 to the original edge cost (Fig. 6(b)).
- Find the least cost path from s to d in the modified graph using the modified Dijkstra's algorithm, and denote it as path p_2 (Fig. 6(c)).
- Remove any edge in the original graph traversed by both p_1 and p_2 (Fig. 6(d)). These are called interlacing edges.
- Identify all path segments remaining after the process of the edge removal (Fig. 6(e)) which forms a pair of disjoint paths from the source node s to the destination node d.

5.3 Algorithm for finding K shortest disjoint path-pairs between nodes in the network

In path protection techniques such as path restoration, or shared backup path protection, there are two sets of candidates needed for provisioning of the working flows and the backup routes. Finding and employing these two sets independently may compromise the protection conditions as the backup and working paths must be disjoint. There is no guarantee that each working route has a disjoint path in the corresponding set of backup routes. In addition, using two different sets of candidates also increases the number of decision variables used in model selection which increases the complexity of both space requirement and computational time. Another approach is to determine one set of candidates in which each candidate is a disjoint path-pair serving as a primary and a backup route. This set would always satisfy the disjointness requirements for protection and reduce the complexity in both time and space of model selections such as Integer Linear Programming (ILP) and Mixed ILP (MILP). By combining algorithms in Sections 5.1 and 5.2, this section introduces an algorithm for finding K shortest (least cost) disjoint path-pairs between two nodes in the networks.

Algorithm 5 K disjoint-paths pairs (KDPPs)

Require: An undirected graph $G(V, S)$ source node s, destination node d, and K, the number of shortest disjoint-paths pairs required.
Ensure: A set of K shortest disjoint-paths pairs.
1: Find a shortest path between s and d, denoted by p.
2: Define the direction of each edge traversed in p from s to d as positive 1.
3: Remove all directed edges on the shortest path p and replace them with reverse direction edges by multiplying the original edge cost with -1.
4: Find K shortest path (least cost) paths from s to d in the modified graph using the K shortest path algorithm (Algorithm 4). Denote them as the set of paths $P = \{p_1, p_2, \ldots, p_K\}$.
5: For each pair of paths $(p, p_i), i = 1 \ldots K$, remove any edge of the original graph traversed by both p and p_i. These are called interlacing edges. Identify all path segments from the rest of edges. These from disjoint-paths pairs, denote as $\{(w_1, r_1), (w_2, r_2), \ldots, (w_K, r_K)\}$.

The two-step and one-step approaches can be used to find the set of disjoint path-pairs. They are simple and computationally efficient. However, since these approaches provision traffic streams sequentially without back-tracking, sometimes no solution may be found even when a feasible solution does exist. In fact, the graph theory approach is more applicable in online provisioning than in offline provisioning. These approaches have been extensively studied in the literature (Patre et al., 2002; Sen et al., 2001; Xin et al., 2002; Zhang et al., 2003). Authors in (Xin et al., 2002) investigated survivable routing based on both the two-step and the one-step approaches, namely Separate Path Selection (SPS) and Joint Path Selection (JPS) respectively. These approaches aim to optimise the network resource utilisation of each traffic connection by minimising the total cost of the primary and backup paths. The performance of these approaches is evaluated for different protection path cost functions. The results have shown that JPS performs substantially better than SPS and also scales very well in terms of the network resource abundance. In addition, JPS can utilise network resources better than SPS. In (Zang et al., 2003), the two-step and one-step approaches are also investigated, but with a different objective function based on the blocking probability. The simulation results have shown that the one-step approach significantly outperforms the two-step approach in this case. The improvement in blocking probability is significant, around 10%-20%, especially when fixed routing is used.

In this section, the one-step approach is used as a sub-function in an algorithm for finding K shortest disjoint path-pairs between any two nodes in the graph. The pseudo code of this algorithm is shown in Algorithm 5. This algorithm is then proved to yield K distinct disjoint-path pairs. Since a path pair found from one step algorithm was proved to be disjoint in (Bhandari, 1999) , it is only necessary to prove that the K found path pairs do not coincide with each other.

Proof. Let $P = \{P_k | k = [1 \ldots K]\}$ be the set of K pairs of disjoint paths yielded from Algorithm 5, where $P_k = \{w_k, r_k\}$ is the kth pair.

Let $P_i = \{w_i, r_i\}$ and $P_j = \{w_j, r_j\}$ be any two disjoint-path pairs yielded from the first shortest path p and paths p_i and p_j $(i, j \in [1 \ldots K])$ respectively. This study proves that P_i and P_j do not coincide.

Since p_i and p_j do not coincide, there exists at least one edge e_k contained in p_i but not contained in p_j. Following this, it is proved that e_k can only belong to either P_i or P_j. If e_k is contained in p then e_k is obviously not contained in P_i. However, since e_k is not contained in p_j, e_k is not an interlacing edge of (p, p_j) and hence e_k is in P_j. Conversely, if e_k is not contained in p, then e_k is in P_i. In addition, since e_k is not contained in both p and p_j, e_k is not contained in P_j. Therefore, P_i and P_j do not coincide and the K found disjoint-path pairs are distinct from each other. □

6. Conclusion

In this chapter we have investigated and discussed the connectivity of graphs in relation to the requirements of protection of traffic demands in practical networks. We have discussed and analysed the complexity of several methods for verifying the connectivity of the physical topology of a network. Diverse routing was then addressed as an important problem in designing network protection, followed by the introduction of three algorithms for finding distinct and disjoint paths. These algorithms address the challenges for network protection design using graph theory.

7. References

Bellman, R. (1958). On a routing problem, *in Quarterly of Applied Mathematics* 16(1): 87–90.

Bhandari, R. (1994). Optimal diverse routing in telecommunication fiber networks, *Proceedings of the 13th IEEE Networking for Global Communications.*, Vol. 3, Toronto, Ont. , Canada, pp. 1498 –1508.

Bhandari, R. (1999). *Survivable Networks: Algorithms for Diverse Routing*, Kluwer Academic Publishers.

Dijkstra, E. (1959). A note on two problems in connection with graphs, *Numerische Mathe-matik* 1: 269–271.

Diestel, R. (2000). *Graph Theory*, Springer-Verlag.

Eppstein, D. (1998). Finding the *k* shortest paths, *SIAM Journal on Computing* 28(2): 652–673.

Grover, W. D. (2004). *Mesh-based Survivable Networks: Options and Strategies for Optical, MPLS, SONET/SDH, and ATM networking*, Prentice Hall PTR.

Habibi, D., Nguyen, H., Phung, Q. & Lo, K. (2005). Establishing physical survivability of large networks using properties of two-connected graphs, *TENCON 2005 2005 IEEE Region 10*, IEEE, pp. 1–5.

Martins, E. D. Q. V., Pascoal, M. M. B. & Santos, J. L. E. D. (1998). The *k* shortest paths problem, *Technical report*, CISUC.

Martins, E. D. Q. V. & Santos, J. L. E. D. (2000). A new shortest paths ranking algorithm, *Investigação Operacional* 20(1): 47–62.

Menger, K. (1927). Zur allgemeinen kurventheorie, *Fund. Math* 10(95-115): 5.

Paton, K. (1969). An algorithm for finding a fundamental set of cycles of a graph, *Communications of the ACM* 12(9): 514–518.

Patre, S. D., Maier, G. & Martinelli, M. (2002). Design of static WDM mesh networks with dedicated path protection, *Infocom*.

Sen, A., Shen, B., Bandyopadhyay, S. & Capone, J. (2001). Survivability of lightwave networks - path lengths in WDM protection scheme, *Journal of High Speed Networks* 10(4): 303–315.

Surballe, J. (1974). Disjoint paths in a network, *Networks* 4: 125–145.

Tarjan, R. (1971). Depth-first search and linear graph algorithms, *Proc. th Annual Symposium on Switching and Automata Theory*, pp. 114–121.

Xin, C., Ye, Y., Dixit, S. S. & Qiao, C. (2002). A joint lightpath routing approach in surviavble optical networks, *Optical Networks Magazine* 3(3): 13–20.

Zang, H., Ou, C. & Mukherjee, B. (2003). Path-protection routing and wavelength assignment (RWA) in WDM mesh networks under duct-layer constraints., *IEEE/ACM Transactions on Networking* 11(2): 248–258.

Zhang, J., Zhu, K., Sahasrabuddhe, L. & Yoo, S. B. (2003). On the study of routing and wavlength assignment approaches for survivable wavelength-routed WDM mesh networks, *Optical Networks Magazine* 4(6): 16–28.

8

Combining Hierarchical Structures on Graphs and Normalized Cut for Image Segmentation

Marco Antonio Garcia Carvalho and André Luis Costa

School of Technology, University of Campinas

Brazil

1. Introduction

The image segmentation task is to divide an image into regions of interest that are suitable for machine or human operations. The machine ability of recognizing and distinguishing objects, or its parts in a scene, is the main goal of computer vision domain. This is a critical issue because the judgement of good or bad segmentation is usually subject to humans.

Extensive studies have been accomplished for image segmentation. The segmentation algorithms are commonly categorized according to the image characteristics borders and regions (Gonzales & Woods, 2000). In the first one, the image is divided based on its discontinuities, *i.e.*, the places where abrupt intensity changes occur. Regarding to the region segmentation, this happen when there are similarities of color or texture, for example, between neighboring pixels. In spite of these categories, the problem is to find a good partitioning of an image among several possible to be achieved.

Several algorithms in image segmentation can be formulated from the partitioning of graphs. This means using graphs as image models or representations and then apply a criterion or methodology in order to split it into subgraphs. The existing literature on graph partitioning is wide, but we are interested on a particular approach, the called Normalized Cut introduced by Shi & Malik (1997).

A graph cut consists of removing edges consistently in order to generate two subgraphs. The Normalized Cut approach is a graph-cut technique based on Spectral Graph Theory responsible for generating balanced subgraphs through the removal of the smallest possible number of edges. The Normalized Cut approach uses concepts by Fiedler (1975) in the manipulation of the second smallest eigenvector of the graph representative matrix as a guide for graph partitioning. The inherent bias of this technique is that balanced partitions for image segmentation cannot be appropriate for some images when small number of partitions is desired (*e.g.*, images with an easily detectible object and an uniform background). A survey of application of Spectral Graph Theory is given by Spielman (2007).

There are several ways of generating the graph model representing the input images. The graph commonly used as input to the Normalized Cut implementation is that one based on the pixel grid similarity. We propose two alternative representations of graphs known as Quadtree and Component Tree similarity graphs. These graphs decompose the image into partitions and thus carry hierarchical information that can be useful on segmentation task.

Another goal is to reduce the computational cost, due to the reduction of the graph size, compared to using the pixel similarity graph.

Finally, we show experiments using Normalized Cut and Quadtree and Component Tree similarity graphs. In addition to the Component Tree, we propose the use of the Reverse Component Tree in order to profit the relevant information in decomposition process of an image. The results are classified using a benchmark provide by The Berkeley Image Database (Martin et al., 2001).

This chapter is organized as follows. Section 2 introduces graph concepts, including our Quadtree and Component Tree based similarity graph. Section 3 presents graph cuts and Normalized Cut theory. Some related works also are described in this section. An overview of the proposed approach is given in Section 4. Experiments in sampled images are done in Section 5 and further comments of the experiments, conclusions, as well as suggestions of future work are done in Section 6.

2. Graph representation

In digital image processing, a graph is commonly used to model digital images (Wilson & Watkins, 1990). In the Normalized Cut segmentation technique the input graph is called *Similarity Graph* (Shi & Malik, 2000), that we explain in details in Section 3.3. In a similarity graph the edge weights should reflect the similarity between the nodes connected by them, and are given by a *Similarity Function*. Here we present the methods for building a similarity graph that we have used in the experiments described in Section 5.

There are several approaches to represent an image as a similarity graph. In the next subsections we present some fundamental concepts on graphs, and four image-graph representation approaches: based on the *Pixel Grid* (Shi & Malik, 2000) , *Multiscale Pixel Grid* (Cour et al., 2005), based on the *Component Tree* (Carvalho, Costa, Ferreira & Cesar-Jr., 2010), and based on the *Quadtree* (Carvalho, Costa & Ferreira, 2010). Approaches based on the *Component Tree*, *Quadtree*, and *Multiscale Pixel Grid* rely on hierarchical structures to model an image as a graph, and provide different segmentation results when compared to the classical non-hierarchical pixel grid approach.

2.1 Basic concepts on graphs

A *graph* is a mathematical structure employed to model or to describe objects and their relationships, *e. g.*, a composition relationship can describe objects and their constituent parts. Let $G = (V, E, W)$ be a non-directed weighted graph; V is a set of *nodes*, E is a set of *edges* $e(i, j)$, $i, j \in V$, and W is a set of *weights* $w(i, j)$, $i, j \in V$. For each edge $e(i, j) \in E$ exists an associated weight $w(i, j) \in W$, that can be represented by a single value or a set of values. Two nodes i and j are *adjacent*, represented by $i \sim j$, if there is an edge connecting i and j. Given a node $i \in V$, its *degree* d_i corresponds to its number of neighbours, and its *strength* s_i (Wilson & Watkins, 1990) to the sum of its edge's weights. A *subgraph* of a graph $G = (V, E, W)$ is a graph $G' = (V', E', W')$ where $V' \subseteq V$, $E' \subseteq E$, and $W' \subseteq W$.

A *path* $\pi = (i_1, i_2, \ldots, i_n)$, $i_n \in V$, is a sequence without repeated nodes where $i_k \sim i_{k+1}$, $k = 1, 2, \ldots, n - 1$. Two nodes are *connected* if there exists at least one path between i and j. In a *connected graph* G all pair of nodes are connected. A *cycle* is a path where $i_1 = i_n$. A *tree* is a connected graph with no cycles.

2.1.1 Graphs and matrices

A graph and its features can be represented using matrices (Wilson & Watkins, 1990). The adjacency between graph nodes can be described by an *Adjacency Matrix* A with size $n \times n$, where n is the number of nodes, $|V|$, of graph $G = (V, E, W)$. The matrix elements $a(i, j)$ are 1 if $i \sim j, i, j \in V$, or 0, otherwise. Similarly, the *Weight Matrix* W, also with size $n \times n$, $n = |V|$, can store the graph weights $w(i, j), i \sim j, i, j \in V$. The *Degree Matrix* D is diagonal with $d(i, i) = d_i$, where d_i is the degree os node $i \in V$. Finally, the *Laplacian Matrix* L is defined as $L = D - A$ for unweighted graphs, or $L = D - W$ for weighted graphs. The Laplacian matrix is commonly used on Spectral Graph Theory (Spielman, 2007).

2.2 Graph based on the Pixel Grid

In this *Pixel Grid-based* image-graph representation each pixel is taken as a graph node, and two pixels within a r distance are connected by an edge. Shi & Malik (2000) have used this approach as the first one for their Normalized Cut technique. This approach have been introduced in the experiments as a landmark for compare results with other approaches.

2.2.1 Multiscale Pixel Grid

The *Multiscale Pixel Grid* graph decomposition algorithm introduced by Cour et al. (2005) works on multiple scales of the image grid to capture coarse and fine detail levels. The construction of the similarity graph is done according to their spatial separation, as in the following Equation

$$W = W_1 + W_2 + \ldots + W_s, \tag{1}$$

where W is a weight matrix that represents a graph composed by independent subgraphs W_s, s corresponds to a scaled pixel grid.

In the Multiscale approach there exists one different radius for each image scale s. Thus, two pixels i and $j \in W_s$ are connected only if the distance between them is lower than a radius R_s. The radii values are a tradeoff between the computation cost and the segmentation result. The Multiscale approach can alleviate this situation by using recursive sub-sampling of the image pixel grid.

2.3 Graph based on the Component Tree

The *Component Tree* (CT) (Carvalho, Costa, Ferreira & Cesar-Jr., 2010) is a hierarchical representation of a digital image after thresholding operations between its minimum and maximum gray values (Mosorov & Kowalski, 2002). There exists a relation of inclusion between components at sequential gray levels in the image, explained below by the partition definition (Carvalho, 2004).

Definition 1. *A partition P of an image I is a set of disjoint regions* $R_i, i \in \mathcal{N}$, *where* $\bigcup_i^n R_i = I$ *and* $R_i \cap R_j = \varnothing, i \neq j$.

A cross-section I_k of an image I is a binary image defined as (Mosorov & Kowalski, 2002; Najman & Couprie, 2006):

$$I_k = \{x \in I / I(x) \geq k\}. \tag{2}$$

In the CT, the *Connected Components* (CC) of all cross-sections are organized in a tree structure. There exists an edge between two connected components CC_{k+1}^i and CC_k^j when $CC_{k+1}^i \subseteq CC_k^j$ (inclusion relation); k is a cross-section identifier, i is a CC in cross-section $k+1$, and j is a CC in cross-section k. The connected component of the first cross-section corresponds to the whole image domain and it is called root. The leaves are the elements of the CT that have no children. A fast algorithm to build the CT is given by Najman & Couprie (2006). Fig. 1 shows a gray scale image I and its five cross-sections I_k. The Component Tree for image I is depicted in Fig. 2(a).

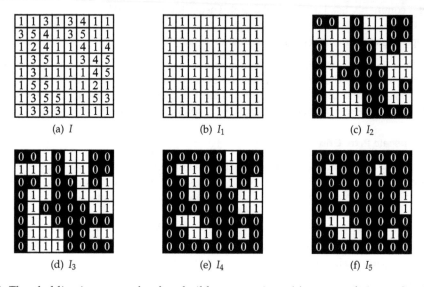

(a) I

(b) I_1

(c) I_2

(d) I_3

(e) I_4

(f) I_5

Fig. 1. Thresholding image graylevels to build cross-sections. (a) a grayscale image I and (b)-(f) its five cross-sections.

When observing the traditional CT model, one realizes that the tree will be composed by only *white* components, i. e., components with value 1. There still information in the cross-sections related to the *black* components. In fact, these components can provide more relevant information in particular cases. Therefore, we have defined the Reverse Component Tree (RCT) (Carvalho, Costa, Ferreira & Cesar-Jr., 2010) where two connected components CC_k^i and CC_{k+1}^j are linked when $CC_k^j \subseteq CC_{k+1}^i$; k is a cross-section identifier, i is a CC in cross-section $k+1$, and j is a CC in cross-section k. Unlike described for the CT case, the roots of the RCT's are formed by the connected components of the last cross-section. Fig. 2(b) shows the Reverse Component Tree for image I, presented in Fig. 1(a).

In order to build a connected similarity graph, we combine Component and Reverse Component Trees. This similarity graph will be used in the graph cut process. First, a connected subgraph $G_k = (V_k, E_k, W_k)$ is created for each cross-section I_k, where the nodes

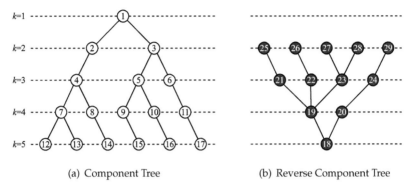

(a) Component Tree   (b) Reverse Component Tree

Fig. 2. Component Tree and Reverse Component Tree from grayscale image presented in Fig. 1(a).

$v_{k,n}$ correspond to the connected components CC_k^n; $k \in \mathcal{N}$ is a cross-section identifier, and $n \in \mathcal{N}$ is a connected component identifier. Given two nodes $v_{k,i}$ and $v_{k,j}$, $v_{k,i} \sim v_{k,j}$ if distance$(v_{k,i}, v_{k,j}) \leq r$, where $r \in \mathcal{R}$ is the connection radius. After built, the k subgraphs are connected by adding the edges from the CT and RCT. The weight of each edge should reflect the likelihood of different connected components. Some of them are difference between areas, density, average gray levels and standard deviation, and Euclidean distance.

One strong characteristic of CT method is the generation of multiple image partitions at once. Because a cross-section I_k corresponds to the whole image, there is one image partition for each cross-section.

Finally, it is important to note that some images can produce a high quantity of connected components, especially in the presence of noise. Therefore, it is useful to apply some pre-processing on the image before starting the CT computation such as, gaussian filter, normalization of the gray values into a smaller range, and merging of identical subsequent cross-sections.

2.4 Graph based on the Quadtree

A Quadtree is a data structure formed from the recursive decomposition of a space (Samet, 1984). In the image processing domain, a quadtree usually maps an image and its regions into a directed acyclic graph (Consularo & Cesar-Jr., 2005).

The decomposition process is simple: The initial region corresponds to the whole image and is associated to the root node; each region in the image should be recursively decomposed into exact four new disjoint regions until they satisfy a defined criterion of homogeneity. Fig. 3 shows a Quadtree decomposition example. The decomposition criterion choice in this case was to exist regions with only one value.

In practice, the regularity of the Quadtree decomposition limits the application of this method to square images with edge sizes 2^n, $n \in \mathcal{N}$. One solution is to relax the regular decomposition, allowing a more suitable number of regions. But the greatest

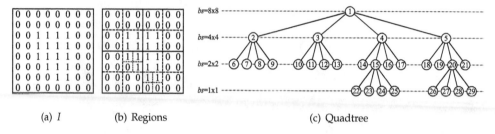

(a) I (b) Regions   (c) Quadtree

Fig. 3. Quadtree decomposition. (a) original image. (b) decomposed regions. (c) corresponding Quadtree.

difficulty in creating the quadtree is the choice of the decomposition criterion. There are different criteria proposed, such as the standard deviation or entropy of image gray levels (Consularo & Cesar-Jr., 2005).

In our quadtree approach (Carvalho, Costa & Ferreira, 2010) for image segmentation, we proposed applying the Canny (Canny, 1986) (1986) filter in the image before the decomposition. This filter has low sensitivity to noise in images. By removing pixels with low gradient and thresholding the resultant ones, this process results in a binary image with border pixels highlighted. This procedure was chosen because:

1. after the filtering, the image results in a binary matrix. Then, facilitating the decomposition criterion definition, that a region should be decomposed when it is not formed entirely by ones or zeros (Samet, 1984);

2. the edge detection operation drastically reduce the size of data to be processed, while at the same time preserves the structural information about object boundaries (Canny, 1986).

In our work (Carvalho, Costa & Ferreira, 2010), the main reason for using a Quadtree image representation was to reduce the similarity graph size. Thus, in this technique the graph is generated using only the regions associated to the Quadtree leaves. Each region is associated to a node and for each region a centroid pixel is defined as the representative pixel. The nodes of the similarity graph are linked together if their representative pixel distances are less than a given radius r, similar to the pixel grid approach. However, it is useful to consider the nodes region sizes to calculate the radius r.

The number of nodes on the similarity graph can be influenced by the choice of the edge detector parameters. The number of regions obtained by the proposed technique will vary according to the image features. Also, the parameters of the edge detection filter can be manually specified, in order to change its sensibility.

3. Graph Cut and Normalized Cut

A graph cut partitions the set of nodes V of a graph $G = (V, E, W)$ into two disjoint subsets A and B, and can be expressed by the following equation (Shi & Malik, 1997):

$$Cut(A, B) = \sum_{u \in A, v \in B} w(u, v), \tag{3}$$

where $w(u,v)$ are the edges removed from G.

This formula indicates the degree of dissimilarity between G_A and G_B, the corresponding subgraphs to the node subsets A and B, respectively. There are a lot of ways to solve the problem of graph cuts. One of them, proposed by Wu & Leahy (1993), solve this problem removing the smallest possible number of edges. However, in some cases this approach can produce isolated nodes. A study about types of graph cut is given by Soundararajan & Sarkar (2001).

A graph cut can be accomplished by means of *Spectral Graph Theory*. In this approach, the matrix eigenvectors that represent graphs are analyzed and used as parameter in order to partition a graph.

3.1 Spectral Graph Theory

The *Spectral Graph Theory* (SGT) studies the eigenvalues of the graph matrices, also called graph spectrum. Algebraic methods used to analyze matrices of graphs are especially effective in treating regular and symmetric graphs (Chung, 1997). The matrices commonly used are adjacency matrices and the SGT establishes a relation between the graph spectrum and the graph features.

The use of graph spectrum information for graph cuts has a great contribution from Donath & Hoffman (1972); Fiedler (1975); Pothen et al. (1990). Fiedler proposed that the second smallest eigenvector v_2 of the Laplacian matrix, also called the Fiedler vector, has in a given row $v_2[i]$ a numerical information about node i, also called the characteristic value of the node. The graph nodes can be partitionated by grouping them according to their value in the Fiedler vector. The commonly way used to group nodes is the characteristic values signals.

3.2 The Normalized Cut

The *Normalized Cut* (NCut) technique (Shi & Malik, 1997) is a theoretic method for graph partitioning. Its goal is to find a balanced cut in a graph, in order to generate two or more subgraphs. This technique solves the problem stated by Wu & Leahy (1993) in their minimum cut criteria for graph cutting. Applying this method for image segmentation is possible with a proper image-graph representation, where the subgraphs obtained from graph partitioning represents the image regions. The NCut in a graph G is calculated by the following equation:

$$\text{NCut}(A,B) = \frac{\text{Cut}(A,B)}{\text{SumCon}(A,V)} + \frac{\text{Cut}(A,B)}{\text{SumCon}(B,V)}, \tag{4}$$

where A and B are the node subsets of subgraphs G_A and G_B, subject to $A \cup B = V$ and $A \cap B = \varnothing$; $\text{Cut}(A,B)$ is defined in Equation (3); $\text{SumCon}(A,V)$ is the total weight of the edges connecting nodes from a subgraph G_A to all nodes in the original graph G; and $\text{SumCon}(B,V)$ is similarly defined to a subgraph G_B.

The optimal NCut is the one that minimizes Equation 4. The problem in minimizing Equation 4 is that it is only trivial for small graphs. For bigger graphs, it has a NP-Complete complexity. Shi & Malik (2000) extended this equation and found a well-known equation in linear algebra called the Rayleigh Quotient. It can be minimized using spectral graph properties of the graph's Laplacian Matrix described by Fiedler (1975), *i.e.*, its minimum value is λ_2, the second smallest eigenvalue (Golub & Loan, 1989).

The graph partitioning is guided by the eigenvector v_2, where each value $v_2[i]$ will represent a graph node i. To split a graph, a threshold value is used and the graph nodes are partitioned in two subsets. The most common threshold values are zero, the median value in v_2 or the one that minimizes the NCut value.

3.3 Similarity Graph

In a *Similarity Graph* the edge weights represent the degree of similarity between the linked nodes. For graphs that represent images, the similarity can be determined by a function of intensity, position and other image pixels features. A measure of similarity regarding the intensity and the position of image pixels is given by (Shi & Malik, 2000):

$$W_{\mathrm{IP}}(i,j) = \begin{cases} e^{-\left(\frac{\alpha^2}{d_p}\right)-\left(\frac{\beta^2}{d_i}\right)}, & \text{if } \alpha_2 < r \\ 0, & \text{Otherwise} \end{cases} \tag{5}$$

where $\alpha = ||P_i - P_j||$ and $\beta = ||I_i - I_j||$ are respectively the distance and the difference of intensity between pixels i and j; r is a given distance (also called graph connection radius); and d_p and d_i are set as the variance of the image pixels positions and intensity. This grouping cue used separately often gives bad segmentations because some natural images are affected by the texture clutter.

The intervening contours is another measure to evaluate the affinity between two nodes by measuring the image edges between their correspondent pixels. The intervening contour similarity function is given by (Cour et al., 2005):

$$W_{\mathrm{C}}(i,j) = \begin{cases} e^{-\left(\frac{\max_{(x \in \mathrm{line}(i,j))}\varepsilon^2}{d_c}\right)}, & \text{if } \alpha < r \\ 0, & \text{Otherwise} \end{cases} \tag{6}$$

where $\mathrm{line}(i,j)$ is a straight line joining pixels i and j and $\varepsilon = ||\mathrm{Edge}(x)||$ is the image edge strength at location x.

These two grouping cues can be combined as (Cour et al., 2005):

$$W_{\mathrm{IPC}}(i,j) = \sqrt{W_{\mathrm{IP}}(i,j)\,W_{\mathrm{C}}(i,j)} + W_{\mathrm{C}}(i,j). \tag{7}$$

3.4 Related work

There is a wide range of recent work in image segmentation using the Normalized Cut technique. The contributions are focused on improving the algorithm performance, others on proposing different image-graph modelling and others on the application of this technique for real-world applications.

In several works, the image model used is the similarity graph built by taking each image pixel as a node. In this case, the node pairs within a given radius r are connected by an edge. This graph will be explained in the next section.

Monteiro & Campilho (2008) proposed the Watershed Normalized Cut, which uses the regions from the Watershed image segmentation as nodes for the similarity graph. The Watershed region similarity graph is either used by Carvalho et al. (2009) for comparison with the primitive pixel affinity graph in yeast cells images segmentation. Ma & Wan (2008) used Watershed based similarity graphs to segment texture images.

The primitive normalized cut enhancement was also studied and applied by many researchers. Cour et al. (2005) proposes a Normalized Cut adaptive that focus on the computational problem created by long range graphs. The authors suggested the use of multiscale segmentations, decomposing a long range graph into independent subgraphs. The main contribution of this technique is that larger images can be better segmented with a linear complexity. Sun & He (2009) proposed the use of the multiscale graph decomposition, partitioning the image graph representation at the finest scale level and weighting the graph nodes using the texture features.

Tolliver & Miller (2006) suggested an improvement of the normalized cut technique. They proposed the use of the k first eigenvectors for graph partitioning as the k-way Normalized cut. The difference is that these eigenvectors modify the edges weight in the graph, resulting in new graph matrices, and the k first eigenvectors are calculated again in the new Laplacian Matrix. The authors proved that this procedure changes the k first eigenvalues to zero. Spectral graph theory concepts about the Laplacian matrix informs that the number of eigenvalues equal to zero shows the number of connected components in a graph. Their algorithm returns these connected components. Cour et al. (2005) suggested another improvement by dividing the graph in scales and processing them in paralell. This approach can segment large images graphs with high conections with linear complexity.

Tao et al. (2008) proposed a new image thresholding technique using the normalized cut. The graph similarity matrix proposed is now based on pixel gray levels, reducing the matrix size and the computational cost. So, a new matrix M is created, where $M(i,j) = \mathrm{Cut}(V_i, V_j)$ with i and j being two given gray levels. Using this matrix, the normalized cut is then calculated to each threshold value, If the normalized cut related to a given threshold value t is below a prespecified value, this threshold value is adequate to separate the objects from the background in this image.

Grote et al. (2007) suggested the normalized cut for extracting roads from aerial images. In their graph, pixels are the graph nodes and the similarity matrix uses contours, hue and color in the image pixels. this approach uses the k-way normalized cut, with k being large to avoid road and non-road pixels mixture. Senthilnath & Omkar (2009) compared this technique with other state-of-art road extraction approach, proving that the normalized cut based technique works better. Other normalized cut aplications in the literature are the noise reduction in images by Zhang & Zhang (2009) and the colour image segmentation by Tao et al. (2008).

4. Normalized Cut segmentation workflow

Given an image I, we build a similarity graph $G = (V, E, W)$, from Quadtree and/or Component Tree representations. The image segmentation process based on Normalized Cut technique can be applied by two distinct methods: recursive Two-way and K-Way. The block diagram presented in Fig. 4 illustrates the complete image segmentation process.

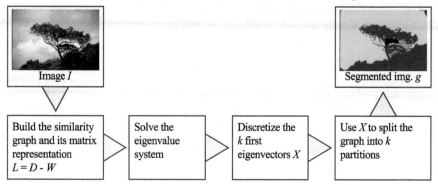

Fig. 4. Workflow of the image segmentation technique based on Normalized Cut.

The similarity graph obtained from Quadtree and Component/Reverse Tree representations is done according to the details explained previously and its parameters are described in Section 5. They have hierarchical information that is exploited in the graph cut process. The eigenvectors of the Laplacian matrix are obtained from the solution of the following equation:

$$(D - W)v = \lambda D v. \tag{8}$$

At this moment, it is possible to use only the second eigenvector v_2 in order to provide two partitions of the graph G and reapply the process to obtain recursively a large number of partitions. A partition should be divided by analysing a specified cut value. This technique is known as Recursive Two-Way NCut. In the other hand, it is possible to discretize the k first eigenvector X, were $X = [v1, v2, \ldots, vk]$ and use them directly to implement the graph partitioning into k desired partitions. This partitioning process is called K-Way Cut and corresponds to the block sequence presented in Fig. 4.

5. Experimental results and discussion

We have performed k-way Normalized Cut segmentation in 100 grayscale test images from *Berkeley Segmentation Benchmark*[1] (Martin et al., 2001). The goal of these experiments is to compare the techniques for building the similarity graph based on *Pixel Grid* (Shi & Malik, 2000), *Multiscale Pixel Grid* (Cour et al., 2005), *Quadtree* (Carvalho, Costa & Ferreira, 2010), and *Component Tree* (Carvalho, Costa, Ferreira & Cesar-Jr., 2010). Fig. 5 show a collection of 9 selected images.

The Berkeley's benchmark rely on human image segmentations to state the segmentation algorithms assertiveness, according to Precision (P) and Recall (R) metrics (Davis & Goadrich,

---

[1] Available at http://www.eecs.berkeley.edu/Research/Projects/CS/vision/bsds/ (last accessed September, 2011).

Fig. 5. Selected images from Berkeley's benchmark. (a) 3096. (b) 21077. (c) 42049. (d) 85048. (e) 97033. (f) 119082; (g) 147091. (h) 167062. (i) 241004.

2006; Martin et al., 2001). Precision is the probability that a pixel marked as a border is in fact a border pixel, and is given by

$$P = \frac{TP}{TP + FP},$$ (9)

where TP is the number of true positives, and FP the number of false positives. The precision P decrease as increases the number of false positives. Recall, also called *hit rate*, is the probability that the border pixels marked by the machine are the same as the border pixels marked by humans, and is defined as

$$R = \frac{TP}{TP + FN},$$ (10)

where FN is the number of false negatives.

These two metrics are summarized in the F-measure (Davis & Goadrich, 2006)

$$F = 2 \cdot \frac{P \cdot R}{P + R},$$ (11)

that is used as the score metric by Berkeley's benchmark to ranking the algorithms effectiveness.

5.1 Experiment setup

The experiments were perfomed according to the workflow presented in Section 4 with k-way method. The connection radius for the *Pixel Grid* graph was $r = 10$ and the edges weights were given by Equation (6). For the *Multiscale Pixel Grid* approach were used one radius for each scale, which were $r_1 = 2$, $r_2 = 3$ and $r_3 = 7$ and the edge weights were given by Equation (6). The *Quadtree* weights were also given by Equation (6), and their radii given as

$$ r_{i,j} = \frac{1}{2} \max(v_i^{\phi x} + v_j^{\phi x}, v_i^{\phi y} + v_j^{\phi y}) + k, \tag{12} $$

where $v_i^{\phi x}$ and $v_i^{\phi y}$ are repectively the horizontal and the vertical diameter of region correspondent to node $v_i \in V$; $v_j^{\phi x}$ and $v_j^{\phi y}$ are similarly defined for node $v_j \in V$; and $k = 10$.

The Component Tree was generated with radius $r = 25$ for the subgraphs. The attributes used to build our CT similarity graph were difference of area, distance, standard deviation of the gray levels and density. For the edges that links the subgraphs, the weights were multiplied by a factor given by the following equation:

$$ f = \frac{NC_{Fi} + NC_{Fj}}{2 + |\,d(i) - d(j)\,|}, \tag{13} $$

where NC_{Fi} and NC_{Fj} are the number of CCs of the cross-sections that has the nodes i and j respectively; and $d(i)$ and $d(j)$ are the degrees of nodes i and j respectively, related to the CT. After the CT segmentation procedure the various partitions generated for each image were converted to a single one by two distinct approaches: manually by selecting the partition that seemed to be the better image segmentation by a criterion of resemblance to the original image; automatically by merging pairs of partitions with highest mutual information iteratively. These approaches sets our experiments on image segmentation based on CT as semi-automatic or automatic procedures.

5.2 Results, discussion, and future works

The Berkeley's benchmark combine the individual scores from all segmentations of each algorithm in a single final score that determine the algorithm overall ranking. In our experiments the first place in the ranking was for the *Multiscale Pixel Grid* approach and the last place was for the automatic *Component Tree*. Fig. 6 show the overall scores obtained for each similarity graph method.

The segmentation scores for some individual images, however, are quite different from the algorithm's overall score. Fig. 7 show the individual scores for each image with each algorithm. We can observe that even the automatic CT approach, which is the worst ranked algorithm, has the highest score to about ten images. A similar result was obtained by the Semi-automatic CT approach. Put together, they account for 20% of the better scores for individual images. Not surprisingly, they also have the highest overall *Recall*, 0.65 and 0.64

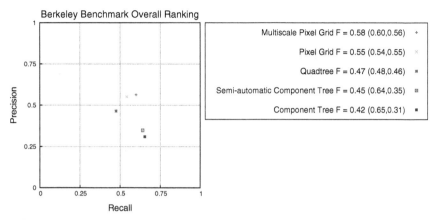

Fig. 6. Algorithms overall ranking. The two values between parenthesis at the right box represent (R,P).

for automatic and semi- automatic methods, respectively, against 0.60 for the best ranked algorithm.

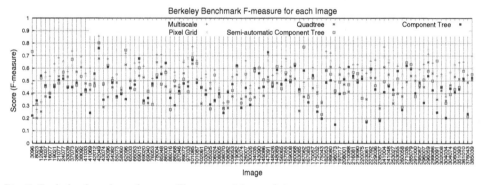

Fig. 7. Berkeley benchmark score (F-measure) for each image.

In the chart shown at Fig. 7 one can yet observe a clear trend for the Multiscale and the Quadtree approaches to follow the scores obtained by the classical Pixel Grid method: Multiscale with a little better, and Quadtree with a little worse scores. We can associate this trend to the fact that these algorithms are using the same similarity function.

Despite these quantitative analisys, it is also important to make a qualitative analisys of the segmentation results. Thus, are shown in Fig 8 one machine segmentation for each image shown in Fig. 5, and its human segmentations.

The computational performance is also an important requirement for the Normalized Cut technique. When Cour et al. (2005) proposed the Multiscale approach to generate the similarity graph, one main objective was to reduce the NCut computational cost. It is also the objective of the Quadtree approach. Fig. 9 show a chart with the time each algorithm took to process the segmentation. There is a strong correlation between the Component Tree

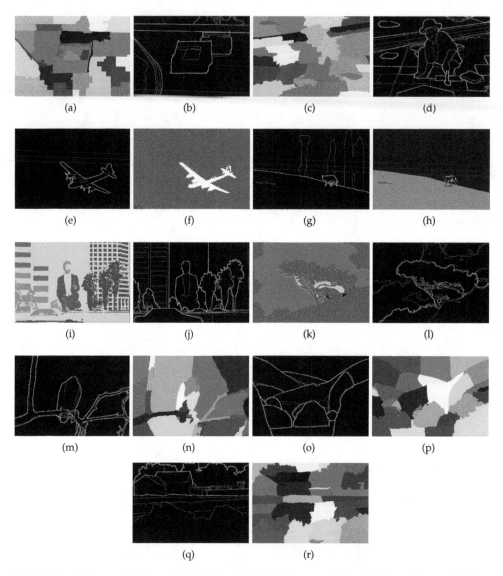

Fig. 8. Selected segmentations. (a) 21077 and (c) 85048 segmentations by Quadtree method. (f) 3096 and (h) 167062 segmentations by Semi-automatic Component Tree method. (i) 119082 and (k) 147091 segmentations by Component Tree method. (n) 42049 and (p) 241004 segmentations by Multiscale Pixel Grid method. (r) 97033 segmentation by Pixel Grid method. (b) 21077, (d) 85048, (e) 3096, (g) 167062, (j) 119082, (l) 147091, (m) 42049, (o) 241004, and (q) 97033, human segmentations from Berkeley benchmark.

and the Quadtree methods. That indicates that they are similarly sensitive to the image's data, but their overall performance are, respectively, the lower and the higher ones. The overall performance of the Multiscale algorithm is higher than the overall performance of the Pixel Grid algorithm, but is lower than the performance of the Quadtree method. Despite the lower performance, only the Component Tree can generate multiple image partitions at once. This problem can be alleviated by implementing the Najman's (Najman & Couprie, 2006) fast algorithm to build the Component Tree.

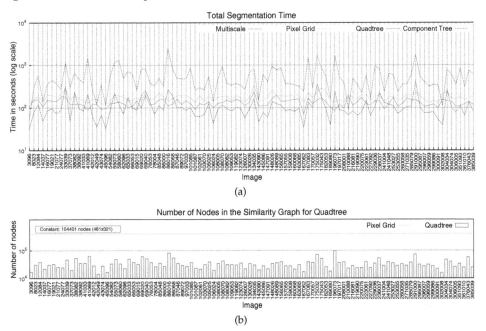

Fig. 9. Algorithms performance. (a) execution time for each image. (b) number of nodes in the similarity graph for Quadtree representation.

The chart in Fig. 9(b) show the number of nodes in the similarity graph generated by the Quadtree approach. Notice that the number of regions in the graph has direct impact to the algorithm's performance.

For future works we see that the Component Tree is the most promising method, despite its worst effectiveness and efficiency. Its implementation can be improved to reduce the computational cost. A more detailed study about the similarity criteria has the potencial of reducing the false positive rate.

There are future works for the Quadtree approach as well. Once the nodes represent regions instead of pixels, the study of texture and other region-based similarity criteria would improve the method effectiveness. It is also reasonable to explore further the hierarchical information of the Quadtree, that would lead the design of a new multiscale approach.

6. Conclusion

In this Chapter we proposed an approach to implement image segmentation based on graph modelling and Normalized Cut technique. Performing graph partitioning by means of Normalized Cut has been widely used in the specific literature. The possibility of generate balanced partitions has shown that this approach is efficient. The proposed similarity graphs, build from the Quadtree and Component Tree structures, proved promising compared to the traditional modelling based on pixel similarity graph. We performed comparisons using two-well established metrics, the Precision and Recall values. An additional aspect to be considered in the proposed graph models is exploring the hierarchical structures of both, Quadtree and Component Tree. Besides, we exploit a little more the Component Tree and proposed the Reverse Component Tree as a way to better represent the information contained in the image cross-sections.

The experimental results accomplished on images from the Berkeley Database show that use regions, or primitive regions to be specific, instead pixels seems to be a better strategy to segment images by Normalized Cut approach. In addition, the new image representation had the advantage of reducing the number of graph nodes and, therefore, improved the algorithm performance.

7. Acknowledgment

This work is supported by FAPESP (São Paulo Research Foundation) – Proc. 2010/14759-0.

8. References

Canny, J. (1986). A computational approach to edge-detection, *IEEE Transactions on Pattern Analysis and Machine Intelligence*, pp. 679–700.

Carvalho, M. A. G. (2004). *Análise hierárquica de imagens através da árvore dos lagos críticos*, PhD thesis, Faculdade de Engenharia Elétrica e Computação- FEEC.

Carvalho, M. A. G., Costa, A. L. & Ferreira, A. C. B. (2010). Image segmentation using quadtree-based similarity graph and normalized cuts, Lecture Notes in Computer Science, v. 6419, pp. 329-337.

Carvalho, M. A. G., Costa, A. L., Ferreira, A. C. B. & Cesar-Jr., R. M. (2010). Image segmentation using component tree and normalized cuts, *Proceedings of the XXIII Brazilian Symposium on Computer Graphics and Image Processing (SIBGRAPI2010)*, Gramado - Brazil, pp. 317-320.

Carvalho, M. A. G., Ferreira, A. C. B., Pinto, T. W. & Cesar-Jr., R. M. (2009). Image segmentation using watershed and normalized cuts, *Proc. of 22th Conference on Graphics, Patterns and Images (SIBGRAPI)*, Rio de Janeiro - Brazil.

Chung, F. (1997). *Spectral Graph Theory*, American Mathematical Society.

Consularo, L. A. & Cesar-Jr., R. M. (2005). Quadtree-based inexact graph matching for image analysis, *Proceedings of the XVIII Brazilian Symposium on Computer Graphics and Image Processing (SIBGRAPI2005)*, Natal - Brazil, pp. 205–212.

Cour, T., Bénézit, F. & Shi, J. (2005). Spectral segmentation with multiscale graph decomposition, *Proc. of IEEE Computer Society Conference on Computer Vision and Pattern Recognition - CVPR'05*, Vol. 2, pp. 1124–1131.

Davis, J. & Goadrich, M. (2006). The relationship between precision-recall and roc curves, *ICML '06: Proceedings of the 23rd international conference on Machine learning*, ACM, pp. 233–240.

Donath, W. & Hoffman, A. (1972). Algorithms for partitioning graphs and computer logic based on eigenvectors of connection matrices, *Technical report*, IBM.

Fiedler, M. (1975). A property of eigenvectors of nonnegative symmetric matrices and its applications to graph theory, *Czech. Math. Journal* 25(100): 619–633.

Golub, G. H. & Loan, C. F. V. (1989). *Matrix Computations*, John Hopkins Press.

Gonzales, R. & Woods, R. (2000). *Digital Image Processing*, Addison-Wesley.

Grote, A. et al. (2007). Segmentation based on normalized cuts for the detection of suburban roads in aerial imagery, *IEEE Proceedings of Urban Remote Sensing Joint Event* pp. 1–5.

Ma, X. & Wan, W. (2008). Texture image segmentation on improved watershed and multiway spectral clustering, *Proceedings of International Conference on Audio, Language and Image Processing - ICALIP*, pp. 1693–1697.

Martin, D., Fowlkes, C., Tal, D. & Malik, J. (2001). A database of human segmented natural images and its application to evaluating segmentation algorithms and measuring ecological statistics, *Proc. 8th Int'l Conf. Computer Vision*, Vol. 2, pp. 416–423.

Monteiro, F. C. & Campilho, A. (2008). Watershed framework to region-based image segmentation, *Proc. of IEEE 19th International Conference on Pattern Recognition - ICPR*, pp. 1–4.

Mosorov, V. & Kowalski, T. M. (2002). The development of component tree for grayscale image segmentation, *Proc. of International Conference on Moderns Problems of Radio Engineering, Telecommunications and Computer Science - TECSET*, Slavsko – Ukraine, pp. 252–253.

Najman, L. & Couprie, M. (2006). Building the component tree in quasi-linear time, *IEEE Transactions on Image Processing* 15(11): 3531–3539.

Pothen, A., Simon, H. D. & Liou, K.-P. (1990). Partitioning sparse matrices with eigenvectors of graphs, *SIAM J. Matrix Anal. Appl.* 11: 430–452.
    URL: *http://portal.acm.org/citation.cfm?id=84514.84521*

Samet, H. (1984). The quadtree and related hierarchical structures, *ACM Computing Surveys* 16(2): 187–261.

Senthilnath, M. R. & Omkar, S. N. (2009). Automatic road extraction using high resolution satellite image based on texture progressive analysis and normalized cut method, *Journal of the Indian Society of Remote Sensing* 37(3): 351–361.

Shi, J. & Malik, J. (1997). Normalized cuts and image segmentation, *Proceedings of the IEEE Computer Society Conference on Computer Vision and Pattern Recognition*, pp. 731–737.

Shi, J. & Malik, J. (2000). Normalized cuts and image segmentation, *IEEE Transactions on Pattern Analysis and Machine Intelligence (PAMI)* 22(8) pp. 888-905.

Soundararajan, P. & Sarkar, S. (2001). Analysis of mincut, average cut and normalized cut measures, *Workshop on Perceptual Organization in Computer Vision*.

Spielman, D. (2007). Spectral graph theory and its applications, *Proc. of 48th Annual IEEE Symposium on Foundations of Computer Science*, pp. 29–38.

Sun, F. & He, J. P. (2009). A normalized cuts based image segmentation method, *Proc. of II International Conference on Information and Computer Science*, pp. 333–336.

Tao, W., Jin, H., Zhang, Y., Liu, L. & Wang, D. (2008). Image thresholding using graph cuts, *IEEE Transactions on Systems Man and Cybernetics Part A-Systems and Humans* 38(5): 1181–1195.

Tolliver, D. & Miller, G. (2006). Visual grouping and object recognition, *Graph partitioning by spectral rounding: Applications in image segmentation and clustering*, pp. 1053–1060.

Wilson, R. & Watkins, J. (1990). *Graphs: An Introductory Approach*, John Wiley and Sons.

Wu, Z. & Leahy, R. (1993). An optimal graph theoretic approach to data clustering: theory and its application to image segmentation, *IEEE Trans. Pattern Analysis and Machine Intelligence*, Vol. 15, pp. 1,101–1,113.

Zhang, L. & Zhang, J.-Z. (2009). Image denoising based on statistical jump regression analysis and local segmentation using normalized cuts, *Acoustics, Speech, and Signal Processing, IEEE International Conference on* pp. 661–664.

9

Study of Changes in the Production Process Based in Graph Theory

Ewa Grandys
Academy of Management in Łódź
Poland

1. Introduction

New areas of knowledge can only be created based on earlier scientific output. To identify possible cognitive gaps, the existing knowledge of management was analysed, which allowed dividing the knowledge hierarchically into three levels of organizational management, enterprise management and production management. The lowest position of the last level in the hierarchy does not make it is less important, especially in the present economic situation of the world that struggles with the impacts of the global crisis.

Graph theory is a field of knowledge offering a broad range of applications. A novel approach was using the theory to build a production management model based on the concept of an inverted tree (with many entries and one exit), as this type of a model reflects the real-life determinants affecting the production of short life cycle goods. The conducted investigation involved clothing companies that produce such goods. Clothing manufacturers deciding to start production invariably expose themselves to considerable risk. This risk can be reduced by obtaining more information on customers' expectations, fashion trends, the conditions of acquiring materials necessary for making the designed models, as well as manufacturer's technical and technological capacity for actually producing the garment. One assumption as made during the investigation was that the analysed product would not be a single design, but a series of garments comprising a fashion collection [6]. It was also necessary to assume a process-based approach to production management that in the opinion of many authors boosts enterprise effectiveness [5, 7]. With these assumptions in mind, the production process was broken down into four main subprocesses (product creation, setting up product manufacturing, manufacturing and sale) and their subdivisions. The presented production management model is based on this breakdown.

According to the research, apparel as a category of the short life-cycle goods is made in a production process (consisting of product creation, the setting up of the manufacturing processes, procurement of intermediate materials and sale) that spans 18 months [6]. In this relatively long time, the actual production circumstances and those predicted at the planning stage are very likely to show significant differences. Why the changes appear will not be explored in the article to keep it concise, but their occurrence will be assumed *a priori* as a fact. This study was designed to investigate the negative effects of the changes delaying the completion of a production process.

It must be remembered that the production cycle for the short life-cycle goods has a pre-determined date when the sale should start. The fact that the delays observed during the research usually resulted from management errors committed for a lack of an appropriate tool aiding production management helped identify a cognitive gap, following which a method for analysing process changes was developed. The method was called a Production Process Control Tree (PPCT).

2. The production cycle model for short life cycle models

2.1 Short life cycle products

The investigation concentrated on production management that in the theory of management deals with providing scientific solutions to production problems. Concentrating on processes „and not on jobs, people, structures and functions makes it possible to watch the main purpose of the process and not its components" [11]. The process-based approach was adopted to create a production cycle model for short life cycle products (SLCP). The term "life cycle" has been „derived from natural sciences, after an analogy between living organisms and products was found. They are too destined to be born, mature, age and die" [1]. The popularity of the term increased with the European Commission's project of 2005 „*The European Platform on Life Cycle Assessment (LCA)*" [25], which extended the notion of life cycle over business and politics. These circumstances increasingly justify the search for regular patterns (cycles, loops, or equilibrium points) that systems tend to [16]. The role of product life cycle was also commented on by P.F. Drucker [3], who stated that it could be used as a tool for evaluating firm's position. Further, R. Kleine-Doepke [12] mentioned life cycle together with the experience curve concept and the results of research conducted under the PIMS (*Profit Impact of Market Strategy*) program, as the third of the factors that contributed to the development of the portfolio planning methods. All these circumstances provided an inspiration for focusing the study on the management of production of the short life cycle products. Such products have one of the below characteristics:

- the length of the selling period depends on product design, and fashion shows cyclical changes,
- the end of the production period is determined by objective constraints on sales (e.g. alternating seasons in the case of clothing),
- the length of production cycle is incommensurate with the selling period.

Typical products whose life cycle is decided by their design are clothing and footwear. Their selling time is correlated with the passing of seasons that together with the seasonal character of fashion trends subject production management to strict discipline. Products manufactured for too long become unsaleable. The manufacturers of durable goods use the positive aspect of fashion to stimulate demand (motor vehicles, interior design articles, etc.).

2.2 Formulation of the research hypothesis

According to M. Marchesnay [19], a theory is constructed via a process consisting of the following stages: formulation of a hypothesis, construction of a hypothesis confirmation procedure, confirmation by means of an empirical test or mathematical-logical evidence, assessment of the confirmation success rate, formulation of conclusions (theoretical/practical). This sequence is criticised by persons advocating the phenomenological approach who believe that understanding the examined phenomena is more important than measuring them. This approach is justified in the case of a heuristic research and scientific procedure. However, M. Marchesnay's procedures are appropriate when the researcher's reasoning is linear. With these reservations in mind, the following hypothesis was formulated:

The management of production of the short life cycle products is a process of determined duration, consisting of four cyclical subprocesses. The cycles are staggered with respect to each other by fixed time intervals.

For the hypothesis to be confirmed an appropriate investigation had to be conducted and its results analysed. The management science proposes in this case a three-stage model of analysis comprising decomposition (into the building blocks), description and integration [23]. The presented algorithm was used at the further stages of the investigation.

2.3 Process decomposition and a description of its components

The process applied to produce garments was divided into four main sub-processes: product development, preparation of production, production and sales. As far as the short life cycle products are concerned, separating product development from all other actions related to the production preparation process is justified. An example of the SLCP is fashion articles whose production is determined by seasonality of sales. Two fashion collections are usually created, i.e. for the spring-summer and autumn-winter seasons.

2.3.1 The garment development subprocess

This subprocess is carried out between 1 February (of the year preceding product release) and 1 August for the spring-summer collection and between 1 August and 1 February of the next year for the autumn-winter collection.

New product designs are created based on the external and internal sources of information. The latter usually account for more than 55% [15]. This rate, however, does not apply to fashion articles. Their originality is assessed by means of criteria laid out in the copyrights. For a season's collection to be created, information has to be collected from several sources:

- design guidelines formulated by fashion creators,
- data provided by a survey of a market segment targeted by the enterprise,
- information about the relevant garment constructions and manufacturing technologies that will be used subject to the availability of appropriate equipment.

A talented designer and the person's ability to assimilate the design guidelines, whose new informative contents alter the knowledge of the trends in fashion development, are a prerequisite for starting the work on a collection. The relevant guidelines can be sought in Paris, Milan, etc., at international fashion shows. As observed, some fashion elements are recurrent; yet, they are never copied, but introduced after some modifications. This justifies the statement that all models of a season's collection are new. Products that are attractive for the buyers (because of their design, quality and price) involve interaction with the customers. Data offered by market surveys are an important source of information, allowing the manufacturer to look at a product from a broader perspective [18]. However, a design that is attractive but priced against the expectations of the targeted market segment will render sales unsuccessful. The responsibility for setting prices rests on an authorised group of persons who take account of the design manufacturing data (material and labour costs, etc.). Design attractiveness is the main criterion affecting the level of prices. A market failure is certain when the season's collection is released too late. This fact demonstrates the importance of the time factor for the production cycle of seasonal products.

2.3.2 The production preparation subprocess

This subprocess is carried out between 1 June and 1 December for the spring-summer collection and between 1 December and 1 June of the next year for the autumn-winter collection.

The production preparation subprocess consists of:

- preparation of the technical and technological documentation,
- procurement of the necessary materials,
- planning and organizing the product manufacturing subprocess.

From the standpoint of the operations to be carried out within the production preparation subprocess, it is not important whether the team of workers responsible for the subprocess will be based in their parent company or at the service provider's site [9]. The technology to be used must correspond to the selected materials and the available machines (either owned or belonging to other party). Modern enterprises prepare the necessary documentation using Computer Aided Manufacturing systems (CAD) that are common in the Polish medium and large-sized manufacturing companies today. The author pioneered the implementation of the first of such systems in Poland in 1987. Other activities related to the preparation of documentation include:

- sequencing the technological operations,
- selecting the machines and pieces of equipment necessary to perform the operations,
- calculating the times necessary to complete each operation.

These documents allow setting up the product manufacturing subprocess, estimating the wages to be paid to the piece-rate workers, the direct production costs, etc.

Material procurement, which is a separate activity, is equally important. The domestic clothing industry buys its materials from suppliers based in the EU (the high end of the apparel market) and in Asian countries (the other segments). The geography of the suppliers usually has a bearing on fabric quality, wearability and price, the latter being a key factor shaping the final price of a product. Fabric patterns are picked by the designers who take into account the designs to be produced, while the procurement personnel is responsible for analysing the submitted offers and making choices that are optimal for the firm. All procurement, although the contract signing process is exposed to the pressure of time, must be carried out prudently. The type of payment (advance payment, pay on delivery, deferred payment) is an important aspect of negotiations concerning the terms of supply. Optimization of the procurement process is vital for enterprise functioning.

The organizational setup of the product manufacturing process must integrate the following areas:

- human (the necessary number of workers with the required skills and available at the right time and place),
- technical (the availability of the machines and equipment),
- material (the optimal quantity of materials stored in the working areas and available when needed).

The above areas decide about the type of the manufacturing system that will be used: an assembly line, an assembly line with sections, sections with synchronised working teams, a flow system or an arrhythmic system [27]. Other factors that influence the production preparation process include the size of an order, the possibility of making several models simultaneously, etc. The organizational efforts must optimize manufacturing times and costs, as well as providing a product of expected quality (in relation to manufacturers' expenses and target buyers' preferences).

2.3.3 The product manufacturing subprocess

This subprocess is carried out between 1 August and 1 February of the next year for the spring-summer collection and between 1 February and 1 August for the autumn-winter collection.

A characteristic feature of garment production is the input of manual labour. In the developed countries, small series of garments containing particular large inputs of manual labour are produced for the high-end market segments. Regardless of the target buyer, garment production consists of two main stages: cutting out garment elements and their assembly. The operations performed at the two stages depend on the organizational plan and technological documentation prepared beforehand.

The first of the two stages comprises the following operations:

- fabric layers are spread along the cutting table to form a stack (in a manual or automated process),
- the layout of the templates to be cut out is transferred on the stack of fabric,
- garment elements are cut out from the fabric with dies, portable oscillating knives, stationary cutting machines, or automated cutting systems,
- adhesive inserts are bonded with the cut-out elements or sections thereof that need stiffening / strengthening,
- the cut-out elements are numbered/marked,
- the cut-out elements are checked for quality,
- the cut-out elements are bundled up (e.g. 50-piece bundles) and stored.

The cutting room setup aims at keeping the cutting machine busy at all times, as the machine is a central piece of equipment that decides about the output of the semi-finished products. The continuous operation of the cutting machine can be ensured via one of two modes. The first of them requires the keeping of spare stacks of laid-out fabric, which increases the amount of work in progress. The second method allows accomplishing a steady workflow by feeding into the machine sections of stacked fabric that are prepared simultaneously on two or several tables. Therefore, the cutting room operations are mainly determined by the available equipment.

The cut-out elements are then transported to the sewing room so that a finished product can be assembled. The garment assembly process is preceded by the following procedures:

-   reorganization of the working team as required by the technical and technological documentation, allowing for the installation of the necessary machines and equipment and the verification of their reliability,
-   cooperation with the Production Preparation Department in making a model ready for production.

All technological operations in the sewing room can be divided into three groups:

-   sewing,
-   adhesive bonding,
-   thermal processing of a product.

Thread joints are the basic technology employed in garment manufacturing. The sewing machines are divided into universal, semi-automated and automated machines. The automated sewing machines are the state-of-the art technology, but garment complexity makes them more appropriate for performing specific jobs or making simple products. Stitching quality is assessed against the in-house standards, as it ultimately affects the quality of a finished product.

Adhesive bonding makes use of a diffusion process that takes place between adhesive particles and the substrate. Adhesive joints are divided into bonded, heated and welded. A popular adhesive bonding method is contact heating. In most cases, two layers of fabric are bonded together with an adhesive. The elements to be bonded are heated on one side and then mechanically pressed together. The bonding process can be applied to a surface (e.g. inserts reinforcing the chest elements of a jacket), linearly (adhesive stitching) or in a dot-like manner. The criterion for the division is the area exposed to bonding. Adhesive bonding is now an important part of the garment manufacturing process, greatly reducing its time.

In terms of technology, the thermal processing of garments can be divided into:

-   forming, i.e. the shaping of garment elements by applying force, temperature and by moistening, is a major process in the production of heavy garments (overcoats), as the shape of their elements depends also on factors other than the construction of particular elements alone;
-   product upgrading by pleating, applying permanent press, as well as surface or local dyeing of finished products in which process the design is transferred from a paper template onto the product by pressing them together;
-   steaming that involves the use of steaming dummies in order to remove minor workmanship defects or to freshen up a product after it has been stored for a long time or cleaned;
-   the interim pressing operations performed at different stages of the assembly process; they have a strong effect on the quality of final pressing and are carried out immediately after the cut-out elements have been joined together, as after attaching the lining such joints become inaccessible;
-   final pressing that determines the ultimate product quality; the operation can be performed with a whole range of specialist machines and pieces of equipment.

Smart clothes are different from the other types of clothing not because of their construction (that can be copied) but due to fabric quality, the quality of the assembly process and thermal processing.

2.3.4 The sales subprocess

This subprocess is carried out between 1 February and 1 August (one year after the product development process was started) for the spring-summer collection and between 1 August and 1 February for the autumn-winter collection.

The Kotler model [14] as applied to the sales subprocess is well known, so it will not be discussed here to make the article concise. Because the scientific output in this field is considerable, it was decided that repeating it would be pointless.

3. Product life cycle and production cycle

The earlier quoted authors suggested that it was possible to present research results as a Kotler curve [14] extended to include the earlier subprocesses. The curve describes a product life cycle as sales in time (fig. 1).

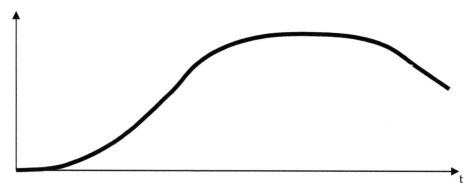

Fig. 1. Product life cycle – Kotler curve
Source: developed after Ph. Kotler, *Marketing Management, 11th ed.*, Prentice Hall 2003, p. 328

According to Kotler [14], the sales curves vary depending on the product (fig. 2):

A. stylish products (architecture, furniture, etc.),
B. fashion articles (clothing, footwear etc.),
C. products with temporary popularity (Rubik's cube, yo-yo, etc.).

The B curve reveals some special character of the fashion articles. The graph is fully adequate when a fashion article is meant as one model of a product. However, the investigation has confirmed that the classical product life cycle approach can be applied to clothing too (fig. 1), if we assume that such products do not represent a single design but a set of designs comprising a seasonal offer, which is in fact the final product of an enterprise. This knowledge encouraged the author to present the investigation's findings graphically as a Kotler curve extended to include the earlier production management subprocesses. Consequently, an original production cycle model for the SLCP was created, approximating the temporal work intensity distribution in the particular components of the production process. According to Waszczyk and Szczerbicki [26], such approximation is rarely treated as model falsification. It is rather assumed that some developments taking place in the economic reality were omitted from the model. The Leśkiewicz [17] accuse the approach of being internally inconsistent, i.e. combining unrealistic assumptions and a strongly accentuated previdistic function of the economy. Nevertheless, Waszczyk and Szczerbicki [26] view things differently: a solid description provided at the stage of constructing an explanatory model lends credence to its prognostic value. The long-time research conducted by the author of this article allows her to confirm this opinion.

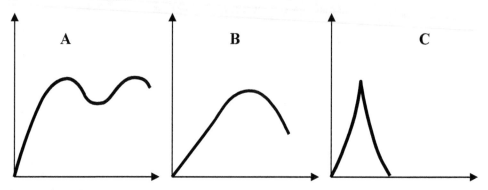

Fig. 2. Life cycles: A – stylish products, B – fashion articles, C – temporarily popular products.
Source: developed after Ph. Kotler, *Marketing…*, op. cit., p. 330

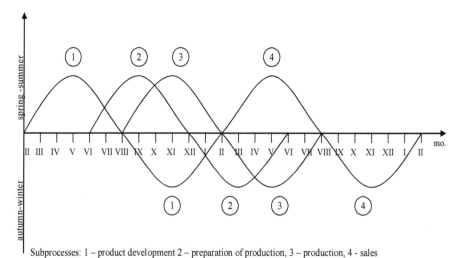

Subprocesses: 1 – product development 2 – preparation of production, 3 – production, 4 - sales

Fig. 3. The SLCP production cycle model.
Source: developed by the author.

The constructed production cycle model (fig. 3) is underpinned by the author's empirical study, the findings of which were verified in the course of her 18 years' long business practice. Although the changing circumstances impeded the correct time measurements, the cyclical character of the production process enabled their verification. For the model to be created „the reality had to be appropriately smoothed out" [12]. It is justified to state now that the presented model sufficiently reflects the actual situation. Author's good knowledge of the research subject is confirmed by the attached description of the production process. All these circumstances together allow concluding that the evaluation [26] of the constructed model' prognostic properties is fully relevant.

In the presented production cycle model, sales are only one of the subprocesses (curve 4). The x-axis separates operations necessary to prepare two clothing collections in a year; these related to the spring-summer season run above the axis, while operations concerning the autumn-winter collection are below it. The curves illustrate the rising and falling intensity of work in the particular subprocesses during the execution of a seasonal collection, as well as determining their acceptable completion times. The curves are shifted with respect to each other by a time interval indicated by the research findings. Although it is possible for the interval to show minor variations, the regular occurrence of a constant time interval is a fact. The interval is:

- four months between curves 1 and 2,
- only two months between curves 2 and 3,
- as many as six months between curves 3 and 4.

Because each of the four subprocesses goes on for as long as 6 months, the production process could be spread over 24 months. The possibility of reducing the time by four months (between curves 1 and 2) and two months (between curves 2 and 3) as demonstrated by the research findings shortens the production process to 18 months. It is also worth noting that the start dates for the spring-summer collection (curve 1 above axis x) and the autumn-winter collection (curve 1 below axis x) are shifted against each other by 6 months. The knowledge of this fact allows optimizing the management of production of seasonal products having a determined manufacturing completion time.

4. A production management model for short life-cycle goods

Production management must always arise from a plan. Every „plan involves a performance imperative. An organization striving to comply with the imperative becomes less flexible and less perceptive of what is going on around it. On the other hand, the increasingly turbulent environment requires organizations to show flexibility, so that unexpected events [...] representing opportunities can be used, while those posing threats avoided [...] "[16]. The changes in the functioning of domestic enterprises are well illustrated by the events that took place at the turn of the 20th c., which demonstrated how the transition of 1989 contributed to the formation of a new economic reality in Poland [7]. After the domestic market was opened, the inflow of imports increased. As a result, stronger competition in the market reduced the range of selling opportunities. These circumstances substantiate an investigation into the ways of improving the effectiveness of production management that provides companies with more favourable market positions.

A production management model for the short life-cycle goods was built following the stages below:

Stage I. Empirical aggregation of actions making up the production process.
Stage II. Using the Altshuller method for verifying causal relationships between particular actions.
Stage III. Making a schedule of the actions and determining their times.
Stage IV. Building the model based on graph theory.

The fact that the author had researched the area for many years helped verify the process constituents and the causal relationships between them were found using the Altshuller method. For the sake of illustration, let us show the applied method with respect to the subprocess representing the creation of a fashion collection.

The evaluation of the attractiveness of clothing designs (5) is the critical action in the string of events presented in diagram 1. The acceptance of the designs means that the next steps can be taken, i.e. the sample room team can prepare a sample of the designed model.

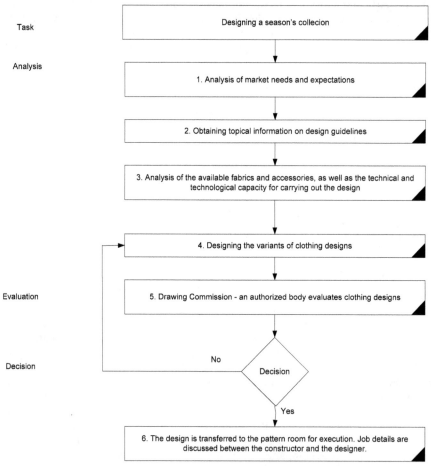

Diagram 1. An algorithm illustrating a fashion collection design process.
Source: developed by the author

Otherwise, the proposed models have to be redesigned. This procedure is repeated until the group of specialists appointed by the enterprise board decides that the outcomes are satisfactory (the number of designs meets the manufacturer's needs).

The Altshuller method applied to evaluate the results of all strings of events allowed scheduling the process correctly. The duration times of the strings were determined empirically and then verified using special catalogues being in possession of every clothing manufacturer. The actual duration times of identical actions may vary between particular manufacturers, as they operate in different technical and technological environments. Therefore, a production management model should follow from an investigation conducted in the concrete enterprise. Because process scheduling is widely discussed topic in the literature [2, 4], for the sake of keeping this article succinct we only wish to note that the schedule produced in the course of this investigation provided a basis for constructing a production management model.

The number of points for starting the model construction corresponded to the number of entry points to the process. The points are the leaves from which each branch consisting of many events (vertices) and of actions (edges) denoting their execution originates. The edges do not provide any information on their duration times, indicating only the sequence of events. At the next stage of tree development, particular branches converge to ultimately form the root, i.e. the final event. With these rules in mind, two assumptions were formulated to build the model:

- the start time of each string of actions depends on the process external and internal determinants,
- the duration of an action is a deterministic value expressed in terms of specific units.

Let us consider whether the second assumption is not a simplification possibly leading to the creation of an unrealistic model. We need to bear in mind that in the case of the short life-cycle goods the time for performing each elementary action can be allowed to deviate from the schedule only to a limited degree, because product selling must start at a predetermined point in time. Therefore, the deterministic time assumption actually relates to some expected time, the length of which is estimated based on long-time experience and many measurements. A starting point for future research could be the adaptation of the model to a situation where the elementary actions have non-deterministic (i.e. described by a random variable) execution times.

The production management model was built along the following lines:

1. Each process action is represented by an edge with a label indicating its duration.
2. There is one vertex for one intermediate state of the process. A vertex is a place where all edges symbolizing actions immediately preceding the state described by the vertex end and where an edge representing an action leading to the next state has its beginning.
3. The tree leaves are equivalent to the initial states of the strings of actions.
4. The root of the tree denotes that the process is complete and the product is ready.
5. Chronologically later process stages are closer to the root than the earlier stages, because the tree is oriented to the root.

6.   Each vertex u can be assigned pairs of numbers $p(u)$ and $q(u)$ denoting the earliest and latest acceptable times of starting actions originating in the vertex.

Let us explain now the exact meanings of the notions and terms used in this article.

The earliest acceptable action start time $p(u)$ is the time when the state u appears, assuming that the preceding actions were performed on schedule.

The latest acceptable action start time $q(u)$ represented by an edge originating in the vertex u allows completing the entire process on schedule, provided that all the following actions are performed on time.

For each vertex u of the production management model, the following inequality exists:

$$p(u) \leq q(u) \tag{1}$$

If the strong inequality $p(u) < q(u)$ is met, then some extra time is available to the state u, which can be used to make up for any earlier delay in the string of actions. The acceptable length of this delay is given by $q(u) - p(u)$ and it is not likely to affect the process completion time, unless the times of the next actions grow longer. However, if the equality $q(u) = p(u)$ takes place, any delay in the string of actions preceding the state u may defer the process completion time. Therefore, the extra time $r(u)$ available to the state u (i.e. the difference between the latest and earliest acceptable start times for an action) can be defined as:

$$r(u) = q(u) - p(u) \tag{2}$$

Based on the above, other assumptions can be formulated:

-     if $r(u) = 0$, the state u will be called sensitive,
-     if $r(u) > 0$, the state u will be called resistant.

To simplify the formulas, a state in the process represented by a vertex (e.g. u) will be treated as identical with that vertex, so the term state u will be used. Analogously, an action originating from the state u and having the state v as its end will be equated with the u–v edge and called the u–v action for the sake of clarity. The time of its execution will be denoted as $t(u$–$v)$. These assumptions allow forming a production management model as a tree. Let us present its portion corresponding to the product creation subprocess (diagram 2).

The designer does his job (4), which is central to the design making subprocess, based on the leaves (actions 1, 2, 3 in diagram 1). The accepted drawings (5) are transferred to the sample room (6). At the same time, visually attractive fabrics are being sought, selected and purchased for the sample room (7–11). The fabrics and the prepared markers (12) are used for making the cutouts (13) that are then assembled with the accessories (14) to make a sample of the model (15). Parallel to that, the abridged model documentation is being prepared, so that the manufacturing costs can be calculated (16–20). The finished models are evaluated for their appearance and functionality (21). The accepted ones are priced and added to the fashion collection (22). If the production process were designed as part of B2B services involving the delivery of corporate clothing (employees meeting with customers shape their employer's image), then body measurements (23-26 in diagram 3) would precede the making of the markers.

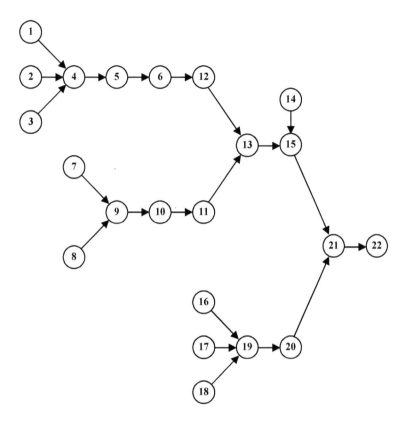

Diagram 2. A production management model – the product creation subprocess.
Source: developed by the author

The decisions giving the green light to the production of particular models (subsequent to consumer focus group, shows at fairs, etc.) initiate the following processes:

- market grading (27–31),
- delivery of the necessary materials (32–37),
- pattern making (38–41),
- preparation of the technical and technological documentation for the sewing room (42–46),
- operational planning (47–51),
- product manufacturing (52–60).

The discussed production management model is illustrated graphically in diagram 3, but without repeating the string of actions that has already been shown in diagram 2 (the product creation subprocess). Let us concentrate our discussion on the product manufacturing subprocess, which is determined by only several vertices. Therefore, the vertex 52 stands for assembling the „job order" (i.e. the putting together of all materials, accessories, the earlier made sample, and the technical and technological documentation for

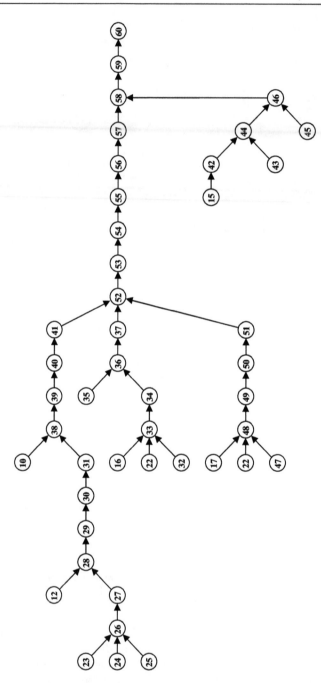

Diagram 3. The production management model.
Source: developed by the author

the manufacturing process), checking the order's completeness and then delivering the complete set to the manufacturing team. The vertices 53–57 represent the cutting room operations (spreading the fabric into stacks, dividing the stacked fabric into sections, making the patterns either automatically or manually, applying the stiffening inserts (if necessary), checking the cut-outs for quality and bundling them together into units of input delivered to the sewing room).

The cutouts and the technological documentation for their assembly meet at the vertex 58 being the string of operations performed in the sewing-room (the operations may vary depending on the model of clothing and the sewing room equipment). For instance, it takes several tens of operations to assemble the cutouts of an overcoat. Given that the construction of the product manufacturing subprocess has already been discussed in another original study by the author [13], it does not seem necessary to repeat it. The vertex 59 is the inspection of the finished product quality and the vertex 60 represents its delivery to the warehouse. Taking into account that the main goal of the investigation was to build a model of production management process, the aggregation of the elementary jobs into sub-processes seems rational.

5. A study of changes in the production process for short-life cycle goods

5.1 A critical path and an Indicator of Process Resistance in the production management model

The critical path of the tree determines the duration of the production process, in the same way as a network's critical path does. By identifying the critical path of the production management model and by calculating the Indicator of Process Resistance (IPR) to change the manager can estimate the degree to which the process completion date is at risk.

Let us present a procedure for finding a critical path or paths of the production management model that do not offer time reserves. Their number provides an indication of the process resistance to changes. The tree has at least one critical path, but every path running between an initial state and a final state can theoretically be a critical one. Then the production process is likely to end later than scheduled, because of changes increasing the duration of any of its constituent tasks. Production processes based on management models with a large number of critical paths are not resistant. The most resistant are processes having only one critical path. This shows that a production management model should be analysed while still being planned, as the range of the model verification options is the greatest then.

Let us present a procedure that was developed to identify the critical path of a production management model shaped as a rooted tree. Let the vertex w correspond to the tree's root (a final event of the production process). Among the states directly preceding the state w there is a state u having a time reserve $r(u) = 0$. If all states coming before the state w had time reserves, then the state w would have a time reserve too. This contradicts the assumption on which model [6] was founded, i.e.:

$$p(w) = q(w), \text{ i.e. } r(w) = 0 \qquad (3)$$

Applying the same reasoning to the state u and its preceding states, we infer that there is some state v preceding the state u, for which $r(v) = 0$. This procedure should be repeated until we discover that a state without a time reserve is one of the initial states (let us call it the vertex a). The path starting at a and ending at w will be called a **critical path**. Accordingly, each state on the critical path is a critical state, particularly the initial state a.

While trying to identify the critical path we may discover that the analysed critical state u is preceded by more than one critical state. Then more than one critical path goes through the state u. This means that **the production management model (for the short life-cycle products)** has as many critical paths as critical initial states. Let us suppose that the model has n initial states containing k critical states. According to the earlier observations, the value of k is within the range:

$$1 \leq k \leq n \tag{4}$$

Consequently, the **Indicator of Process Resistance** to unexpected changes can be calculated as follows:

$$\text{IPR} = \frac{n-k}{n} = 1 - \frac{k}{n} \tag{5}$$

where: IPR – Indicator of Process Resistance to unexpected changes,
n– the number of the initial states (tree leaves),
k– the number of the critical states among the initial states.

IPR's extreme values are obtained for $k = n$ and $k = 1$. They define the range of values for the created indicator, i.e.:

$$< 0; 1 - \frac{1}{n} > \tag{6}$$

When every path in the production management model is critical, then $k = n$ and the IPR = 0. In the model with a single critical path $k = 1$ and the IPR is close to one, because it is calculated as:

$$\text{IPR} = 1 - \frac{1}{n} \approx 1 \tag{7}$$

It can be assumed that such defined **IPR's value measures the process resistance to changes that delay its completion.** The indicator can be used for assessing the production management model's design, as well as its actual performance (each time it has been modified). When the IPR is 0, then the timely completion of the production process is very much at risk, should any change extending the duration of any of its tasks occur. This means that the IPR is a synthetic measure of process resistance to changes, regardless of the stage they affect.

5.2 The PPCT as a method of investigating changes in the production process

A production management model (an inverted tree) provides information on the tasks, their duration and relationships. This aggregate knowledge allowed developing a method that:

-   controls the duration of the process-related tasks,
-   monitors changes affecting task duration,
-   allows adjusting the model, when the changes delay the production process end date.

Let us present the **Production Process Control Tree** (PPCT) method, which was specifically developed for the short-life cycle goods. Let us note that the occurrence of some state u can be delayed with respect to the time $p(u)$, because of one of two reasons:

- one of the states directly or indirectly preceding the state u, e.g. a, occurs later than $p(a)$,
- one of the tasks preceding the state u, e.g. a-u, stretches over a longer period than the scheduled time $t(a$-$u)$.

where: $p(u)$ – the earliest moment of commencing the task originating in the vertex u,
$p(a)$ – the earliest moment of commencing the task originating in the vertex a,
$t(a$-$u)$ – the time for performing the task between the vertices a and u.

The delayed occurrence of the state u may make shift the production end to a later date. This is certain to happen, when u is a critical state. In a general case, **a delay will only take place when the delayed occurrence of the state u affects the commencement of the nearest critical state situated on the path linking the state u and the tree root.** However, the delay of the state u may also be "absorbed" by the time reserves of the states following u.

For the production management model to monitor changes, we need to find the times $p(u)$ and $q(u)$, representing the earliest moment of commencing each task and the latest moment of ending each task, respectively. According to the literature, the following rules can be applied to find the duration of the tasks:

- a deterministic rule for tasks carried out according to the company's own rules that explicitly prescribe task's deadline or duration,
- a probabilistic rule for tasks of duration determined empirically by an experienced expert.

The knowledge of the rules for determining the duration of tasks and time reserves played an important role in developing the PPCT method. Each time reserve indicates the length of time by which a task can extend without the production management model having to undergo adjustment. Considering that a production process is shaped by many variables, the constructed model has to be dynamic, enabling even some modifications to the schedule in case the end date of the production process becomes uncertain.

Let us suppose that we have a production management model shaped as an inverted tree. The duration of particular tasks and the moments when the initial states of tasks should start are also known. The duration of the task transforming the state a into b is denoted as $t(a$-$b)$ and the earliest moment when the state u can commence is denoted as $p(u)$. Let us create an algorithm for this production management model to find $p(a_i)$, $q(w)$ and $q(a)$. The values will be used when the production management model will have to account for the impacts of variables affecting the production process. So:

- $p(a_i)$ is the earliest allowed moment when the task originating in the vertex a_i should commence; it will be calculated according to formula (8) below,
- $q(w)$ is the latest allowed moment of ending the final process activity w; it will be calculated according to formula (11) below,
- $q(a)$ is the latest allowed time when the task originating in the vertex a should end; it will be calculated according to formula (13) below.

Let us use this method for calculating the time $p(u)$ for the vertex u, which is not a tree leaf (i.e. an initial state). In the considered tree, the offspring of the vertex u are the vertices that come immediately before it. Then:

$$p(u) = \max \{p(a_i) + t(a_i - u)\} \quad \text{for } i = 1, \ldots, k \tag{8}$$

where: $p(u)$ – the earliest moment when the task originating in the vertex u should commence; $p(a_i)$ – the earliest moment when the task originating in the vertex a_i should commence, $t(a_i - u)$ – the duration of the task between the vertices a_i and u.

The above **rule helps to find the moments $p(u)$, first for the states directly following the initial states and then recurrently for all vertices of the tree representing a production management model.**

Let w be a state equivalent to the tree root and $p(w)$ the actual end date of the production process. If T is the scheduled end date, then the process will end as planned when:

$$p(w) \leq T \tag{9}$$

If otherwise, the product will not be ready on time. This problem can be handled by adjusting the process, which entails some restructuring of the production management model. Continuing the earlier procedure, we determine the moment $q(u)$, i.e. the latest allowed time when the state u should commence. Naturally, the relationship:

$$q(w) = T \tag{10}$$

still holds.

Building on the earlier assumption about the management of production of short life-cycle goods, we can write that:

$$p(w) = q(w) = T \tag{11}$$

where: $p(w)$ – the earliest allowed moment when the task originating in the vertex w (the tree root) should commence,
$q(w)$ – the latest allowed when the task w, being the final task of the production process, should end,
T – the scheduled process end.

If the equality $p(w) = q(w) = T$ does not take place, then the condition (11) can be met by introducing a dummy state w and assuming that:

$$p(w') = T \text{ and } (w - w') = T - p(w) \tag{12}$$

The time $t(w - w')$ shows the shift in the product completion date, thus providing information on the observed change's effect on the production process. According to assumption (11), **the vertex w ending the production process (i.e. the tree root) is a critical state.** So, the moment $q(u)$ of the state u is determined and the state a precedes the state u towards the root. Then $q(a)$ can be defined using the following equation:

$$q(a) = q(u) - t(a - u) \tag{13}$$

where: $q(a)$ – the latest allowed moment when the task originating in the vertex a preceding the vertex u should end,
$q(u)$ –the latest allowed moment when the task originating in the vertex u should end,
$t(a-u)$ – the duration of the task between the vertices a and u.

The value on the right hand-side of equation (13) is determined precisely, as there is only one edge that starts at a and ends at u. Consequently, the formula needs neither a minimum operator nor a maximum operator. The formula (13) allows recurrent determination of the time $q(u)$ for each vertex of the tree, which property is utilised in the PPCT method to monitor changes.

Changing production circumstances may extend the amount of time that is needed to end the process. This threat must result in an immediate correction of the model data. In other words, changing circumstances should generate warnings about a possibly delayed product completion date.

The PPPCT method we propose for investigating production changes compares the duration of each task with its scheduled time. Let us assume that a task a–x was performed in time $t'(a$–$x)$, which extended beyond its scheduled time $t(a$–$x)$ by n units. Then we have:

$$t'(a\text{–}x) = t(a\text{–}x) + n \tag{14}$$

Let us also assume that the path linking the state x and the tree root successively goes through the states $y_1, y_2, ..., y_k$, so the path is given as $x - y_1 - ... - y_k - w$. We additionally assume that all tasks preceding the state a were performed on time, meaning that the activity a–x started at the moment $p(a)$. Then, one of the three possibilities takes place:

$$1° \quad p(a) + t(a\text{–}x) + n \leq p(x), \text{ or} \tag{15}$$

$$2° \quad p(x) < p(a) + t(a\text{–}x) + n \leq q(x), \text{ or} \tag{16}$$

$$3° \quad q(x) < p(a) + t(a\text{–}x) + n. \tag{17}$$

where: $p(x)$ – the earliest moment when the task originating in the vertex x should commence,
$q(x)$ – the latest moment the task originating in the vertex x should end,
$p(a)$ – the earliest moment when the task originating in the vertex a preceding the state x should commence
$t(a$–$x)$ – the duration of the task between the vertices a and x,
n – the number of units by which the duration of the task a–x has been extended.

Let us explore now the meaning of the three situations and find the algorithms to deal with them.

Should the first case occur, the production process end date runs no risk of being delayed, because the earliest moment when the task y_1 (initiation of the activity x) should start takes place after the length of time allocated to activity a–x elapses. So the model of the process does not need any modifications, but replacing the time $t(a$–$x)$ by $t'(a$–$x) = t(a$–$x) + n$. In this case, the production manager does not have to be informed about an event if occurred, as its influence was neutral.

In the second case, the production process is not exposed to any direct threat to its timely completion, because the state x has the time reserve $r(x) = q(x) - p(x)$. The task a–x will end not later than $q(x)$, being the latest allowed moment for the task x--y_1 to commence. In this situation, the tree requires the following modifications:

i. the time $(a$–$x)$ has to be replaced with $t'(a$–$x) = t(a$–$x) + n$,

ii.   each state on the path $x - y_1 - ... - y_k - w$ has to be assigned a new commencement time using the formula (8) and some obvious changes have to be made to the formula symbols (u has to be replaced with the right vertex name and a_i with its preceding vertices).

The production manager has to be notified of the changes, but no action is required to prevent the production process from running late.

Two things need to be raised at this point: 1) the moments $p(x)$, $p(y_1)$, $p(y_2)$, ..., $p(y_k)$ have to be recalculated, because the changes may have transformed the resistant states into critical ones, thus affecting the process structure; 2) the moments $q(x)$, $q(y_1)$, $q(y_2)$, ..., $q(y_k)$ do not change, because the scheduled process end date $T = p(w) = q(w)$ that depends on them remains the same.

In the third case, the production process end date will be exceeded, because the state x needs more time to end than its time reserve $r(x)$ allows. Hence, the model has to be modified as follows:

i.   the time $t(a-x)$ has to be replaced with $t'(a-x) = t(a-x) + n$,
ii.   each state on the path $x - y_1 - ... - y_k - w$ has to be assigned a new commencement time using the formula (8) and some obvious changes have to be made to the formula symbols (u has to be replaced with the right vertex name and a_i with its preceding vertices),
iii. the process end date T has to be replaced with $T' = p(w)$, assuming at the same time that $q(w) = p(w)$, where the time $p(w)$ represents a new process end date calculated at step (ii),
iv.   now the latest allowed moments of starting tasks that have not been carried out yet have to be determined, their duration of the tasks remaining the same.

The steps (i) ÷ (iv) **readjust the model of the process**. The production manager has to be notified of the situation to decide about the next steps after analysing the new model. The manager may choose to shorten the sequence of activities $x - y_1 - ... - y_k - w$ by introducing organizational, technical or technological improvements. If the intervention is effective and, for instance, the time $t(y_i - y_{i+1})$ becomes shorter by m units, then the steps (i) ÷ (iv) should be repeated, with the time $t(y_i - y_{i+1})$ at step (i) being replaced by $t(y_i - y_{i+1}) - m$.

Because decisions on taking actions causing dynamic adjustment of the model usually increase product manufacturing costs, the company board has to grant the production manager an appropriate scope of authority. Otherwise, the PPCT method enabling interactive management of production processes will not be as effective as it can be. If the duration of the longest path of the tree ensures following the intervention that the production process will end as scheduled, then the process is continued. If otherwise, the production manager has to notify the Board (or another relevant body) of the situation, which may choose to discontinue the production process (after estimating the losses) or to carry on.

6. Conclusion

The presented operational algorithm allows concluding that the empirical verification of the production model for the short life cycle products confirms the hypothesis of the determined duration of the production process. The process consists of four subprocesses that run cyclically and are staggered with respect to each other by constant time intervals. Because the main building material of the management science is the inductive methods that enable drawing general conclusions from empirical research [22], the presented modelling results can be assumed to have a scientific value. The generated model of the product manufacturing cycle is central to company management. Its importance goes beyond the possibility of synchronising the subprocesses alone. The prolonged production of the SLCP shortens the available selling period, and in the extreme case it can completely ruin the sales. This situation is caused by the seasonal character of sales, the time of the year when a clothing article can be worn, etc.

A production management model based on an inverted tree concept (graph theory) allows an innovative, graphical representation of a management process applied to the production of short life-cycle goods. Although an inverted rooted tree has not been used in management theory so far, there are good reasons for constructing it, because it can help:

- identify the sensitive graph routes, i.e. those determining process duration,
- develop a unique measure of process resistance to change (PRI) enabling an immediate evaluation of the production management model for the designed production process or after each process modification caused by changes arising during its execution,
- develop a method for analysing changes in the management of production of short life-cycle goods that allows its user to have active control over the process.

The amount of material discussing the above issues is quite extensive, so it will be presented in other articles that will be published in this periodical.

All computations (necessary to design the original tree, to determine its longest path and to readjust the production management model) can be performed in real time, once an appropriate computer software has been developed. The need for the management process to integrate technology, organizational issues and IT, regardless of the supervisory, stimulating role of the process manager, was accentuated by B. Nogalski [20]. The software should be built around the aforementioned algorithm that enables study of changes in the production of the short life-cycle goods. The software helps implement the method we propose in business practice. Making decisions under the pressure of time is a key problem that companies manufacturing short life-cycle products have to resolve. Modern production management methods, including the PPCT, combined with IT tools are the only ones that make it possible to:

- analyse a production process and its changing circumstances on an on-going basis,
- make decisions in real time to offset the negative impacts of changing process circumstances,
- minimize the losses a company may incur should it decide to discontinue the production process, because every step forwards generates unnecessary costs (augmenting the losses).

The proposed method for analysing changes in the production process can improve the effectiveness of companies making short life-cycle goods that function in turbulent environments. According to the *Global Trends 2025* report prepared by the National Intelligence Council, such environments are becoming the norm today [21]. The report reveals not only problems, but also opportunities that arise from unexpected changes creating new economic realities.

7. References

[1]   Ayres R.U., Ayres L., Rade I., *The Life Cycle of Cooper, Its Co-Products and By-products*, Kulwer Academic Publishers, Dordrecht 2003

[2] Burchard-Korol D., Furman J., *Zarządzanie produkcją i usługami*, Wydawnictwo Politechniki Śląskiej, Gliwice 2007
[3] Drucker P.F., *Natchnienie i fart, czyli innowacja i przedsiębiorczość*, Wydawnictwo Studio Emka, Warsaw 2004
[4] Durlik I., *Inżynieria zarządzania. Strategia i projektowanie systemów produkcyjnych*, Agencja Wydawnicza Placet, Warszawa 2007
[5] Grajewski P., Nogalski B., *Potencjalne źródła niesprawności w organizacji procesowej*, [in:] M. Romanowska, M. Trocki (ed.), *Podejście procesowe do zarządzania*, Wydawnictwo SGH, Warszawa 2004
[6] Grandys E., *Production Cycle Model for Short Life-Cycle Models*, "Fibres & Textiles in Eastern Europe" 2010, No. 1
[7] Grandys E., *Impact of External Determinants on the Functioning of Polish Clothing Manufactures*, "Fibres & Textiles in Eastern Europe" 2010, No. 3
[8] Grandys E., *Productions Management Model for Short Life-Cycle Goods*, "Fibres & Textiles in Eastern Europe" 2010, No. 4
[9] Grandys E., Grandys A., *Outsourcing – an Innowation Tool in Clothing Companies*, "Fibres & Textiles in Eastern Europe" 2008, No. 5
[10] Gutenberg Th., *A Review of Improvement Methods in Manufacturing Operations*, "Work Study" 2003, Vol. 52, No. 2
[11] Hammer M., Champy J., *Reeingineering the Corporation – Manifesto for Business Revolution*, Nicholas Brealey Publishing, London 1994
[12] Kleine-Doepke R., *Podstawy zarządzania*, Wydawnictwo C.H. Beck, Warsaw 1995
[13] Kołacińska E., *Zastosowanie teorii grafów w kierowaniu zespołami szyjącymi*, „Odzież" 1977, nr 9
[14] Kotler Ph., *Marketing Management. 11th ed.*, Prentice Hall 2003
[15] Kotler Ph., Amstrong G., *Principles of marketing*, 4th ed., Prentice Hall, London 1989
[16] Krupski R., *Innowacje w planowaniu strategicznym*, [in:] H. Bieniok, T. Kraśnicka (ed.), *Innowacje zarządcze w biznesie i sektorze publicznym*, Wydawnictwo Akademii Ekonomicznej w Katowicach, Katowice 2008
[17] Leśkiewicz I., Leśkiewicz Z., *Zarys metodologii ekonomii (część II i III)*, WNUS, Szczecin 1999
[18] de Luca L.M., Atuaehbe-Gima K., *Market Knowledge Dimensions and Cross-Functional Collaboration: Examining the Different Routes to Product Innovation Performance*, "Journal of Marketing" Vol. 71, January 2007
[19] Marchesnay L., *L`economie et la gestion sont – elles des sciences? Essaie d`epistemologie*, „Cahier de l`EFRI" Universite Montpelier 2004, Vol. 11, No. 1
[20] Nogalski B., *Restrukturyzacja procesowa w zarządzaniu małymi i średnimi przedsiębiorstwami*, Oficyna Wydawnicza Ośrodka Postępu Organizacyjnego Spółka z o.o., Bydgoszcz 1999
[21] National Intelligence Council, *Global Trends 2025: A Transformed World*, Washington, U.S. Government Printing Office, November 2008, www.dni.gov
[22] Sudoł S., *Nauki o zarządzaniu. Węzłowe problemy i kontrowersje*, Wydawnictwo TNOiK, Toruń 2007
[23] Szczerbicki E., *Management of Comlexity and Information Flow In Agile Manufacturing*, [in:] A. Gunasekaran (ed.): *The 21st Century Competitive Strategy*, Amsterdam 2001
[24] Szczerbicki E., Jinadasa P., *Modelling for Performance Evaluation In Complex Systems*, „Systems Analysis, Modelling, Simulation" 2000, Vol. 38
[25] *The European Platform on Life Cycle Assessment*, http://lca.jrc.ec.europa.eu
[26] Waszczyk M., Szczerbicki E., *Metodologiczne aspekty opisowego modelowania w naukach ekonomicznych*, Zeszyty Naukowe Politechniki Gdańskiej, Gdańsk 2003
[27] Więźlak W., Elmrych-Bocheńska J., Zieliński J., *Odzież. Budowa, własności i produkcja*, Wydawnictwo Instytutu Technologii Eksploatacji – PIB, Radom 2009

10

Applied Graph Theory to Improve Topology Control in Wireless Sensor Networks

Paulo Sérgio Sausen, Airam Sausen and Mauricio de Campos
Master's Program in Mathematical Modeling, Regional University of Northwestern Rio Grande do Sul State (UNIJUI)
Brazil

1. Introduction

This chapter presents solutions for computing bounded-distance multi-coverage backbones in wireless networks. The solutions are based on the (k,r)-CDS problem from graph theory for computing backbones in which any regular node is covered by at least k backbone members within distance r, offering a variable degree of redundancy and reliability.

Advances in lower power-consumption processors, sensor devices, embedded systems and wireless communication have made possible the development and deployment of Wireless Sensor Networks (WSN).Such networks have been used, for instance, in safety and military applications for the purpose of monitoring and tracking geographic boundaries. In industrial applications, WSN may be used for automating manufacturing processes, for monitoring building structures, and in environmental systems for monitoring forests, oceans and precision farming Akyildiz, Su, Sankarasubramaniam & Cayirci (2002).

A WSN is usually composed by a large number of small nodes, called *sensor nodes*, each one having a processing unit, a radio transceiver and an antenna for wireless communication, one or more sensor units (e.g., temperature, movement), and a power unit usually equipped with a low capacity battery. Due to its limited power resources, and because batteries cannot be easily replaced, nodes are built out from power saving components.

To save energy,both approaches apply the partial or total turn-off of some node units. The rule of thumb is to keep active only those units (or components) necessary for performing the sensor network tasks. However, it is not an easy task to decide which nodes should sleep and which should be active at any given time, because these decisions strongly depend on the application running on top of the network. It is also undesirable to keep nodes inactive for too long, because it can impact the network *Quality-of-Service* (QoS).

Topology control through the construction of backbones can offer better support for broadcasting and routing of data packets. The total or partial turn off of nodes not comprising the Backbone constitutes the main motivation for employing such structures in WSNs. Reduced power consumption, and longer network lifetime, are potential benefits when applying Backbones to a network of sensors, because only a subset of nodes need to be active at any time to support basic network services.

In contention based medium access protocols using a single channel, the number of collisions of packets increases as we increase the number of competing nodes. By reducing the total number of active nodes, backbones can potentially reduce the end-to-end delay among sensors and sinks, and possibly extend the network lifetime due to the reduced energy consumption. In addition to that, it might be possible to provide better Quality of Service (QoS) for WSNs.

The basic criteria of QoS in WSN (i.e., *area coverage*) can still be guaranteed by not totally turning off the sensor nodes. Given that the radio is the most expensive element in terms of energy consumption Margi et al. (2006), by just turning it off and leaving the other components active we can still manage to save on energy.

In Ad Hoc networks Alzoubi et al. (2002); Bao & Garcia-Luna-Aceves (2003a), and more recently in WSNs Dai & Wu (2005); Paruchuri et al. (2005), the computation of Backbones has been considered through the computation of some variant of Connected Dominating Set (CDS). A set of nodes in the network is considered a Dominating Set (DS) if all nodes in the network belong to this set or, otherwise, are adjacent (i.e., neighbors) to at least one DS node. A DS is said to form a CDS when the graph induced by the DS is connected (i.e., the DS induces a Backbone for the network).

The problem of computing a CDS of minimum cardinality for any arbitrary network topology, a Minimum CDS (MCDS), constitutes an NP-complete problem Garey & Johnson (1978a), and requires knowledge of the entire network topology. Distributed solutions for computing approximations to the MCDS problem have been proposed in the literature Chen & Liestman (2002); Das & Bharghavan (1997). Even though such solutions aim at reducing the total number of dominating nodes, it has been shown that in wireless networks there is a tradeoff between redundancy and reliability Basu & Redi (2004). Too much redundancy is not desirable, but too little may compromise the network connectivity. This is of special concern to WSNs, because nodes are more vulnerable to failures due to the environments in which they operate.

Dai e Wu Dai & Wu (2005) proposed a distributed solution for computing a k-connected k-dominating Backbone for wireless networks. Their approach combines multiple domination and the k-vertex connected property, which guarantees that a CDS remains vertex connected even when removing up to $k - 1$ nodes from the backbone. The shortcomings of this approach have to do with the required high degree of redundancy in the network topology.

In this chapter we propose the first centralized and distributed solutions for computing *bounded-distance multi-coverage backbones* in WSNs. This means that any sensor node is covered by multiple backbone members within a bounded-distance. To guarantee these properties, the (k,r)-CDS mechanism is employed for computing the Backbone. The multiple domination parameter, k, defines the minimum number of backbone nodes covering any regular sensor node. The bounded-distance parameter, r, defines the maximum distance to k backbone nodes for any other sensor in the network. The centralized solution provides an approximation to the optimum solution, and it is used as a lower bound when evaluating the performance of the distributed solution. The distributed solution is source-based in the sense that usually the base-station (or sink) is the focus of attention in a WSN.

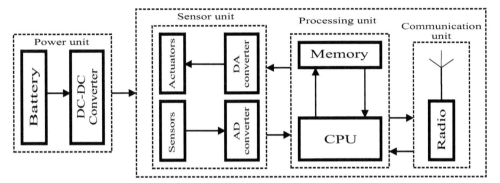

Fig. 1. Sensor node: basic structure

2. Wireless Sensor Networks

A Wireless Sensor Network may well contain hundreds or thousands of small autonomous elements called sensor nodes, and each sensor can feature a large variety of sensors (e.g., temperature, speed, acoustic, seismic). In many cases, nodes are randomly spread over remote areas, making it difficult to perform any maintenance to the nodes. Hence, a node remains live while it has enough battery capacity for its normal operation, and the network lifetime strongly depends on the remaining capacity of the nodes in the network. A sensor node has a few basic components (see Figure 1) Akyildiz, Su, Sankarasubramaniam & Cayirci (2002); Raghunathan et al. (2002) as follows:

- *Power Unit*, usually a battery, which acts as the power source for all node's components;
- *Sensor Unit* that contains a group of sensors and actuators;
- *Processing Unit* which includes a microprocessor or a micro-controller;
- *Communication Unit* which consists of a short range radio for wireless communication.

3. Domination in graph theory

An undirected graph $G = (V, E)$ consists of a set of vertices $V = \{n_1, \ldots n_k\}$, and a set of edges E (an edge is a set $\{n_i, n_j\}$, where $n_i, n_j \in V$ and $n_i \neq n_j$). A set $D \subseteq V$ of vertices in a graph G is called a *dominating set* (DS) if every vertex $n_i \in V$ is either an element of D or is adjacent to an element of D Haynes et al. (1998). If the graph induced by the nodes in D is connected, we have a *connected dominating set* (CDS). The problem of computing the minimum cardinality DS or CDS of any arbitrary graph is known to be NP-complete Garey & Johnson (1978b).

A variety of conditions may be imposed on the dominating set D in a graph $G = (V, E)$. Among them, there are *multiple domination*, and *distance domination* Haynes et al. (1998). *Multiple domination* requires that each vertex in $V - D$ be dominated by at least k vertices in D for a fixed positive integer k. The minimum cardinality of the dominating set D is called the *k-domination number* and is denoted by $\gamma_k(G)$. *Distance domination* requires that each vertex in $V - D$ be within distance r of at least one vertex in D for a fixed positive integer r. In this case, the minimum cardinality of the dominating set D is called the *distance-r domination number*, and is denoted by $\gamma_{\leq r}(G)$.

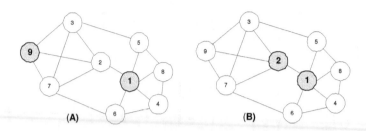

Fig. 2. *Dominating set* examples (gray nodes are *dominating*): (A) Dominating Set (DS); (B) Connected Dominating Set (CDS).

Henning et al. Henning et al. (1996) have presented some bounds on the *distance-r domination number* $\gamma_{\leq r}(G)$. They show that, for an integer $r \geq 1$, if graph G is a connected graph of order $n \geq r + 1$, then $\gamma_{\leq r}(G) \leq \frac{n}{r+1}$. An algorithm that computes a *distance-r dominating set* within the established bounds is also presented.

The (k,r)-DS problem is defined Joshi et al. (1993) as the problem of selecting a minimum cardinality vertex set D of a graph $G = (V, E)$, such that every vertex u not in D is at a distance smaller than or equal to r from at least k vertices in D. The problem of computing a (k,r)-DS of minimum cardinality for arbitrary graphs is also NP-complete Joshi et al. (1993). Figure 2 (A) show example the Dominating Set and Figure 2 (B) show the Connect Dominating Set example.

Joshi et al. Joshi et al. (1993) have provided centralized solutions for solving the (k,r)-DS problem in *interval graphs* (IG). A graph G is said to be an interval graph if there is a one-to-one correspondence between a finite set of closed intervals of the real line and the vertex set V, and two vertices u and v are said to be connected if and only if their corresponding intervals have a nonempty intersection. Even though the solutions presented by Joshi et al. Joshi et al. (1993) are optimal, IGs are limited to very simple network topologies.

4. *Clustering* and topology control

In wireless sensor networks (WSN), nodes must coordinate among themselves because they cannot assume any fixed infrastructure (e.g., access points). Broadcasting of signaling messages is the underlying mechanism for coordination, and the broadcast can target a portion of the network (e.g., gathering neighborhood information), or the entire network (e.g., discovering routes on demand).

Coordination in WSN includes operations such as neighborhood discovery, organization of nodes (i.e., topology control and clustering), and routing. Examples of organization of nodes include the location of services, computing an efficient backbone for the broadcasting of signals, and routing of data packets.

Organization of nodes can be proactive or on demand. While operations to build such structures require broadcasting of signaling messages, these structures make broadcast operations scale to much larger portions of the network. That is, the hierarchical structure functions as a backbone, on top of which broadcasting can be performed more efficiently.

A virtual hierarchy is also possible. In this case, nodes are organized not depending on their physical location, but based on some other criteria. For example, the virtual topology could be comprised only by nodes speaking a given protocol. In this case, a virtual link exists between any pair of nodes whenever they are talking to each other using the specified protocol.

In some cases, the predefined hierarchical address of each node reflects its position within the hierarchy Tsuchiya (1988). In heterogeneous networks (i.e., networks composed of nodes with different capabilities in terms of energy, bandwidth, processing power, and transmission power), not only the physical location but also the resources of each node can be used as a criteria for deciding the role of each node within its sub-structure.

There are two broad categories of hierarchical architectures in ad hoc networks: *clustering* Chen et al. (2004), and *topology control based on hierarchies* Bao & Garcia-Luna-Aceves (2003b); Li (2003). These architectures can be employed to extend the network lifetime Bandyopadhyay & Coyle (2003); Chen et al. (2001); Younis & Fahmy (2004), achieve load balancing Bao & Garcia-Luna-Aceves (2003b), and to augment network scalability Heinzelman (2000); Li (2003).

With *clustering* Chen et al. (2004), the substructures that are condensed in higher levels are called *clusters*. For each cluster there is at least one node representing the cluster, and this node is usually called a *cluster-head*. Cluster-heads act as leaders in their clusters, providing some service to their members. As an example, a cluster-head could be an access point to the outside network, or it could be a *sink* for collecting information from a group of sensors (cluster members) in a wireless sensor network Akyildiz, Weilian, Sankarasubramaniam & Cayirci (2002).

Topology control and clustering are closely related problems. While the former defines a physically connected *backbone* of the network (i.e., the backbone nodes are connected, and they cover all nodes in the network), the latter constructs a *virtual backbone* (i.e., the set of cluster-heads do not necessarily compose a connected component of the network, even though they cover all nodes in the network).

Hierarchical structures are used for broadcasting, transmission of control messages, and routing data packets. The selection of nodes for the backbone will have a direct impact in the overall performance of the protocol. In most cases, dominating nodes are selected using some distributed dominating set algorithm, and if a backbone is desired, auxiliary nodes are selected to connect the dominating nodes.

5. Centralized (k, r)-CDS mechanism

In this section we shows the centralized solution for computes a (k,r)-CDS for any connected networks. The network can be abstracted as a graph $G = (V, E)$ where the set V includes all nodes (i.e., vertices), and the set E includes all links (i.e., edges). The centralized solution for the (k, r)-CDS problem requires that the entire network topology be known. The centralized solution applies a *greedy* heuristic usually employed to optimization problems when there is the need for defining a group of candidates by optimizing the value of a given metric. The centralized mechanism guarantees that all nodes in the network, with the exception of those that comprise the Backbone, are covered by k members of the Backbone within distance r. It is assumed that all nodes have an unique identifier (i.e. *id*) and they know the whole network topology.

While carrying out the algorithm, colors are ascribed to nodes to reflect their states during the selection process. To begin with, all nodes are colored **White**. A node is colored **Gray** when it becomes a candidate to the Backbone. Once a node joins the Backbone, the node is colored **Black**. The heuristic chosen for selecting nodes for the (k, r)-CDS is described as follows:

1. Initially, all nodes in the network are colored white.
2. The node with the largest r-hop neighborhood (i.e. $u \in V$ with maximum value for $|N_r^u|$) is colored black. Ties are broken lexicographically by considering the highest *id*.
3. Next, all nodes adjacent (i.e. neighboring nodes) to the black node are colored gray.
4. The gray node with most white neighbors is then selected and colored black. Ties are broken by choosing the highest *id*. This procedure continues until the conditions in item 6 are satisfied.
5. In the absence of any white nodes, the node with the most gray neighbors is considered as the next member of the Backbone.
6. The process comes to an end when all regular nodes have been covered by at least k members of the Backbone (black nodes) within distance r; i.e., when all nodes in the network, except those that are members of the Backbone, have met the (k,r)-CDS specification.

5.1 Example

Figure 3 shows the computation of a (2,2)-CDS for a particular network topology. Black nodes are members of the Backbone. For this configuration, nodes that do not belong to the Backbone are covered by at least two nodes of the Backbone at most two hops distant. Initially, all nodes are colored white, and have their degrees calculated (Figure 3(A)). For example, the node with id 3 displays the marking $(7, -, 0)$, which indicates that it has seven neighbors within two hops (i.e., $r = 2$), and it is not covered by any black nodes yet (indicated by the third parameter set to zero).

The node with the largest 2-hop neighborhood (i.e., $u \in V$ with maximum value of $|N_2^u|$) is colored Black (Figure 3(B)). At this point, there is a tie among nodes $0, 2$ and 5. In this case, the node with the largest ID (i.e., node 5) wins. Following that, the neighboring nodes to node 5 are colored gray, while its r-hop neighborhood is accounted for this new Backbone node (i.e., all nodes within distance 2 from node 5 are covered by this node).

Node 0 is the Gray node covering the most White neighbors (nodes 4 and 6), hence it is chosen for the Backbone and colored Black. Therefore, all its neighbors are colored Gray (Figure 3(C)). The remaining gray nodes hold only one White neighbor. Once node 0 has been selected, all its r-hop neighbors (i.e., nodes $2, 3, 4, 6, 7, 8, 9$) have their parameters updated to reflect coverage by node 0 (Black nodes do not need to be covered, for they are part of the Backbone).

The process goes on by selecting the Gray node that has the most White neighbors (Figure 3(D)), and in this case node 9 is selected for untying purpose. After this last selection, the Backbone is complete, and all network nodes are covered by at least two Black nodes within distance 2. The Backbone is then formed by nodes 0, 5, and 9.

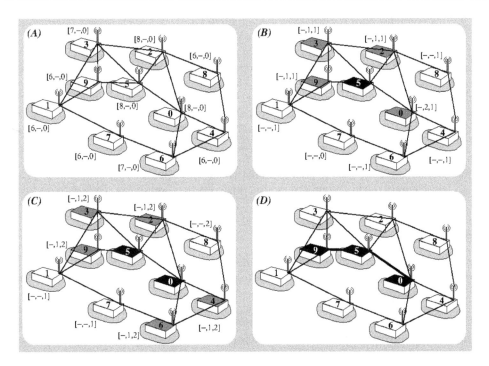

Fig. 3. Exemple (2,2)-CDS applying the centralized solution

5.2 Formal analysis

A WSN is modeled as a graph $G = (V, E)$, where V is the set of sensor nodes, and E is the set of wireless links. Each node in V is associated to its Euclidean coordinates. A link between any pair of nodes u and v exists (i.e., $(u, v) \in E$) if and only if the Euclidean distance between the pair of nodes is smaller than or equal to their transmission range (assumed to be the same for all nodes in V).

A set $V' \subseteq V$ forms a CDS for network G if all nodes in $V - V'$ are adjacent to least one node in V', and the sub-graph $G[V']$ induced by V' is connected.

Lemma 1. *Connectivity is preserved by augmenting a connected component with any node adjacent to the connected component.*

Proof. This can be proved by induction on the cardinality of the connected component. The base case is when the connected component is composed by just one node. Now consider a connected component with cardinality $n - 1$. A node adjacent to any node in the connected component augments its cardinality to n, and keeps the component connected because of the adjacency property. □

Theorem 1. *The centralized mechanism correctly computes a (k, r)-CDS Backbone for any arbitrary connected graph $G = (V, N)$.*

Proof. To account for node's coverage, k sets are used, $D_0, D_1, D_2, ..., D_k$, where D_i represents the nodes covered by at least i nodes from the Backbone within distance of r hops. Initially, all nodes of G are in D_0. The inclusion of a new member to the Backbone can promote one or more nodes from D_i to D_{i+1} for all $i < k$. The nodes elected for the Backbone are immediately inserted into D_k. The first Backbone node, b_0, is the node with the largest r-hop neighborhood in the graph. Whenever a tie occurs, the node with the largest identifier (i.e., **id**) is selected. In each iteration, a new backbone node, u, is selected among the neighboring nodes to the Backbone (i.e., N_B) such that the selected node has the largest number of uncovered neighbors (i.e., $\forall\ n \in N_B$ we have that u maximizes $\sum_{i=0}^{k-1} |D_i \cap N_1^n|$). The connectivity property is guaranteed by Lemma 1. The process ends when all nodes are covered, or all nodes belong to the Backbone (i.e., $D_k = V$). Assuming networks with a finite number of nodes, n (i.e., $n = |V|$), the number of iterations is at most n; therefore, the computation ends within a finite period of time.   □

6. Distributed (k, r)-CDS mechanism

Unlike the centralized solution, the distributed (k,r)-CDS solution does not require the information about the whole network topology. All nodes are required to know only their r-hop neighborhood. The (k,r)-CDS extends the (k,r)-DS mechanism Spohn & Garcia-Luna-Aceves (2006) proposed for ad hoc networks. The (k,r)-DS mechanism is used for the construction of bounded distance multi-clusterhead clusters (i.e., each regular cluster member is covered by several clusterheads within a bounded distance), but it does not connect the clusterheads among themselves (i.e., that is why it is a DS).

As for the centralized solution, the distributed mechanism computes a (k,r)-CDS of the network. However, the computation of the Backbone is accomplished distributively. In addition to that, to adhere to the characteristics of sensor networks, the construction of the Backbone stems from a particular node which could be the Base Station (BS) in the network.

We assume that any node in the network could be the BS. Moreover, there could be more than one BS. However, in this work we consider a single BS per network, which is randomly chosen among all nodes in the network. It is also assumed that each node has an unique identifier, and they have knowledge about their r-hop neighborhood. The mechanism is carried out in two phases as described as follows.

In phase 1 the BS initiates the process by sending an Information Message (IM) to their neighbors reporting the Distance to the Base Station, dBS, initially set to zero by the BS itself. On receiving this message, a node updates its distance to the BS and announces its dBS to its neighbors through a new IM message. If a node gets more than one notification message, any message announcing a shorter distance to the BS triggers a new IM message so that any neighbor can eventually learn about a shorter path to the BS. Considering that all transmissions are reliable, after a finite period of time, and possibly after many retransmissions, all nodes in the network learn their shortest distance to the BS.

The second objective of this phase consists of obtaining the r-hop neighborhood topology, which requires r rounds of messages exchange. In each round, the nodes broadcast information about their known neighbors, and their respective distances. After r rounds, all nodes come to know their neighbors within distance r. By piggybacking the information about each node's distance to the BS (i.e., their dBS), at the end of this phase nodes also learn the distance from each r-hop neighbor to the BS.

Phase 2 starts immediately after the completion of phase 1. It is during this phase that nodes elect k nodes from their r-hop neighborhood to become part of the Backbone. After that, the node announces its elected nodes throughout its r-hop neighborhood. The election is based on the information gathered during phase 1.

The BS node actually starts the election process by electing the k nodes closest to itself (i.e. those with the smaller dBS in the neighborhood). Any ties are broken choosing the node with the largest degree, and persisting the tie the node with the largest ID wins. The BS creates and transmits to all its neighbors a message called Election Message (EM) carrying a list called Backbone Members (BM) with the k elected nodes.

Upon receiving an EM message, a node performs - just once - the selection of its Backbone Members. The criteria are similar to those used by the BS with the restriction that only nodes belonging to the Backbone, or those adjacent to it (i.e. neighbors to Backbone members), can be elected. This restriction guarantees the Backbone connectivity property(i.e. the creation of a CDS), and it reduces the total number of Backbone members because nodes already in the Backbone are likely to cover multiple nodes in the neighborhood.

After a node elects its Backbone members, the BM list in the EM message is updated to reflect any changes to the list, and the message is then transmitted to its neighbors. In addition to that, a Notification Message (NM) is sent to all (if any) new elected nodes (i.e., nodes that were not listed in the original BM list). The election process continues until all nodes in the network have elected their Backbone members. At any stage, a node that has already carried out its election and receives another EM message will just discard it. [1]

6.1 Example

Figure 4 shows the computation of a (2,2)-CDS for a particular WSN. As previously, Black nodes represent Backbone members. Initially, node 0 is selected as the BS, and it is colored Black (Figure 4(A)). Given that $r = 2$, all nodes are familiar with their neighborhood up to 2 hops. The first field in the marking process (shown close to each node in the Figure) identifies the dBS value. The second field indicates whether a node belongs to the Backbone or it neighbors a Backbone member (value 1), otherwise it is set to value 0. The third field indicates the **Degree** of a node.

Using the information obtained during phase 1, the BS selects its two members for the Backbone (Figure 4(B)). Notice that the BS selects itself (because it is a Backbone member by default) and another node from its r-hop neighborhood. The list of elected nodes is transmitted to the BS neighbors through an EM message.

The first elected node embodies the BS itself since $dBS = 0$ and because it starts the whole process. Node 5 is the second elected node for it exhibits the largest id while node 2 ties in Degree and distance (i.e., dBS) with node 5. The election of nodes 0 and 5 reflects on the marking of nodes $2, 3, 4, 6$ and 9, demonstrating that they are either members or neighbors of

---

[1] To handle periodical elections, one could use an election identifier in the EM message to allow nodes to know when they should re-elect their Backbone members.

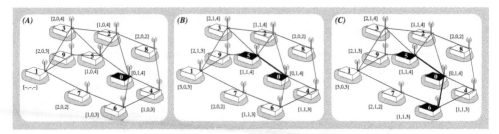

Fig. 4. Exemple (2,2)-CDS applying the distributed solution

a Backbone member. On the sequence, all BS neighbors perform their elections. All nodes with $dBS = 1$ (i.e. the nodes 2, 4, 5 and 6) elect nodes 0 and 5 as members of the Backbone, triggering a new EM message to be sent to their neighbors.

For nodes two hop distant from the BS (i.e., $dBS = 2$), we have that nodes 3, 8 and 9 elect nodes 0 and 5, and node 7 elects nodes 0 and 6 (Figure 4(C)). The marking of node 7 is updated demonstrating that this node is a neighbor of a Backbone member. Node 1 which has $dBS = 3$ receives the EM message and elects nodes 5 and 6 as members of the Backbone. This ends the election since all neighbors of node 1 (which is the node farther away from the BS) have already completed their election. The Backbone is then composed by nodes 0, 5 and 6.

6.2 Formal analysis

To prove the correctness of the distributed (k,r)-CDS mechanism, we have to show that it is *safe* (i.e., the algorithm computes a (k,r)-CDS of the network), and that it is *live* (i.e., it completes within a finite period of time).

Lemma 2. *Phase one of the distributed solution has message complexity of $O(n \cdot r)$, where n is the number of nodes in the network and r is the distance parameter.*

Proof. During each round, nodes send messages to all their one-hop neighbors. Phase one takes r rounds. Assuming a network of n nodes, phase one requires $n \times r$ messages to complete. Therefore, the message complexity of phase one is of order $O(n \cdot r)$.   □

Lemma 3. *After r rounds of successful transmission of message m, the message is propagated up to r hops away from the originating node.*

Proof. This can be proved by induction on the distance d from the node starting the process, n_0. The base case is when $d = 0$, meaning that the starting node itself knows the message it created to be propagated throughout the network. Now consider a node u at distance $r - 1$ from n_0. Once node u retransmits message m to all its one-hop neighbors, the message eventually reach any neighbor r hops from n_0.   □

Theorem 2. *The distributed solution correctly computes a (k, r)-CDS Backbone for any connected graph $G = (V, E)$ during phase two.*

Proof. We assume that any node $i \in V$ knows its r-hop neighbors (i.e., N_r^i), as well as their distances to the base station i_0 (Lemma 3). The base station, i_0, is selected as the first backbone node. On its turn, node i_0 selects for the backbone the $k - 1$ nodes closest to itself (i.e., first it tries to select among its neighbors within distance one, then, if necessary, within distance two, and so on until $k - 1$ nodes are selected). The selected nodes repeat the process, taking as candidates to backbone members those nodes from their r-hop neighborhood which are closer to the base station, i_0, and also requiring that any candidate must be adjacent to or already a member of the backbone. The latter requirement, guarantees the connectivity of the Backbone (Lemma 1). After node i_0 has selected its $k - 1$ backbone members, the base station announces the list of elected nodes to its neighbors, which in turn repeat the process by performing their own election. The election is carried out just once by all nodes in the network (duplicate announcements are just discarded). Considering that messages take a finite period of time to propagate throughout the network, the whole election takes a finite period of time to complete. Given that any node must elect k members among its r-hop neighborhood, and that any elected node must be adjacent to nodes already in the Backbone or members themselves, the connectivity and coverage properties are guaranteed. □

7. Performance analysis

The centralized and distributed (k,r)-CDS mechanisms are compared through extensive simulations using a customized simulator. Network topologies are created based on the *unit disk graph* model Clark et al. (1990). According to this model, any pair of nodes, A and B, are said to be connected if the Euclidean Distance between their coordinates is smaller or equal to R, the transmission range of any node. We assume $R = 15m$ for all nodes (i.e., a homogeneous networks). Nodes are randomly placed over the terrain, and only connected topologies are taken into account. For each configuration, we gather results for 30 trials.

Nodes	Diameter	Degree
200	11,7±0.7	11,8±0.6
400	10,9±0.4	24,2±0.7
600	10,6±0.5	36,6±0.8
800	10,3±0.5	49,0±0.9
1000	10,1±0.3	61,4±0.9

Table 1. Parameters for Scenarios 1

For the simulations we assume an ideal MAC protocol with no collisions, ensuring that all transmissions are successful. The same assumption is made in related work Spohn & Garcia-Luna-Aceves (2006); Wu & Dai (2003). All topologies evaluated in the simulations are assumed to be static (i.e., no mobility). Because there is no known optimum solution for the (k,r)-CDS problem, the proposed centralized solution is used as a lower bound when evaluating the performance of the distributed solution. The simulations encompass a series of experiments changing the distance parameters (i.e., r) and the coverage parameters (i.e., k). All experiments are repeated for 30 trials corresponding to different network topologies.

To gather representative statistical samples, two scenarios are considered. Table 1 presents the average standard values for the **Diameter** (i.e., the largest shortest distance between any pair

Nodes	Diameter	Degree
200	11,9±0.7	12,2±0.6
425	18,0±0.5	11,7±0.5
750	23,2±0.6	12,0±0.0
1185	29,4±0.7	12,0±0.0
1720	35,4±1.0	12,3±0.5

Table 2. Parameters for Scenarios 2

of nodes) and the **Degree** (i.e., average number of one-hop neighbors) over 100 samples for first scenario.

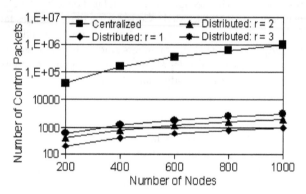

Fig. 5. Control message overhead

In the first scenario, the network diameter stays almost unchanged as the average degree increases with denser networks. As for the second scenario (see Table 2), the network density is kept the same by increasing the terrain size for larger networks. As a consequence, the average degree remains almost the same but the network diameter increases (i.e., the networks are sparser compared to the ones from the first scenario).

To compare the two proposed heuristics, two performance metrics are evaluated: *control message overhead*, the total number of control messages exchanged for gathering topology information (centralized mechanism), or for the execution of the two phases in the distributed mechanism; *total number of Backbone nodes*, the total number of nodes that each mechanism selects to form de Backbone.

Figure 5 shows the results for the message overhead regarding the first scenario. Results for the second scenario are omitted because they are similar to those from the first scenario. Considering a network composed of n nodes, the centralized mechanism incurs $O(n^2)$ messages exchange due to the topology dissemination process. In the distributed mechanism, the control overhead depends on the distance parameter r, because the nodes need to obtain information regarding the r-hop neighborhood. Given that all nodes participate in this process, it incurs a $O(n \cdot r)$ message complexity.

As expected, the centralized mechanism presents the best results for the first scenario (Figure 6). As the distance parameter r increases, the performance of the distributed

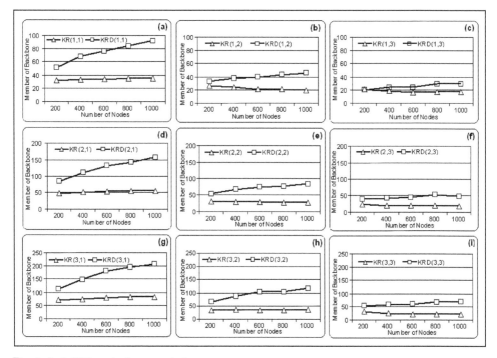

Fig. 6. (k,r)-CDS results for scenario 1

mechanism gets closer to the performance of the centralized mechanism (Figures 6(c), (f), and (i)). The best results are achieved for networks with 200 nodes (Figure 6(c)), when the centralized mechanism elects 20.8 nodes in average, whereas the distributed mechanism elects 21.4 nodes in average. The difference between the two approaches becomes more prominent for larger networks (e.g., networks with 1000 sensor nodes). As the network gets denser, the distributed mechanism elects more nodes. Because nodes elect k nodes among their r-hop neighborhood it is likely that more candidates satisfy the election requirement (i.e., nodes already in the backbone or adjacent to any backbone node), incurring on more redundancy when compared to sparser networks. For the future work we intend to apply some pruning mechanism to reduce redundant backbone members.

For the second scenario, we have chosen to keep node density steady while varying the network diameter. The simulation results (Figure 7) show an increase in the total number of Backbone nodes for both mechanisms. This is due to a better distribution of sensor nodes (i.e., sparser networks). As expected, the centralized mechanism presents better results for most experiments. However, the two heuristics compare to each other in the $(1, 2)$-CDS and $(1, 3)$-CDS configurations (Figure 7 (b) and (c)). As noticed in the first scenario, as the distance parameter r increases, the two approaches present similar results. On the other hand, the difference between the performance for both solutions reduces compared to the first scenario.

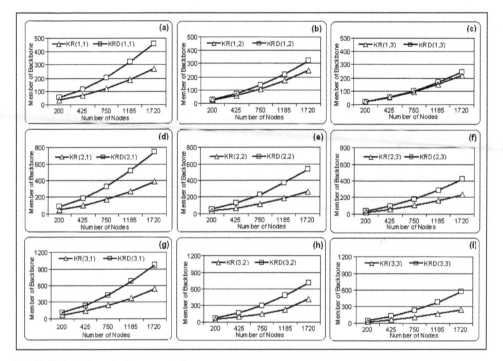

Fig. 7. (k,r)-CDS results for scenario 2

8. Conclusion

We presented the first centralized and distributed solutions for computing *Bounded-Distance Multi-Coverage Backbones* in Wireless Sensor Networks (WSNs). Backbones have the (k,r)-CDS properties, which include the minimum number, k, of nodes in the backbone covering any regular node, as well as a bounded-distance, r, to the covering nodes.

Given the characteristics of WSNs, redundancy is desirable. Therefore, any mechanism requiring redundancy (e.g., data gathering, routing) could take advantage of topologies with the (k, r)-CDS properties. By limiting the distance to the backbone, it could well translate to bounded-delays when accessing the backbone. In any situation, the solutions presented in this paper provide new basis for designing protocols for WSNs.

We compared the centralized and distributed mechanisms through extensive simulations using a customized simulator. Given that the centralized solution is unsuitable for WSNs, because of the incurred control overhead, it is used as a lower bound when evaluating the performance of the distributed solution. It is shown that even though the distributed solution builds larger backbones, it does not incur on much control overhead.

As future work, we are planning applying a pruning mechanism to the distributed solution during the election process. In face of that, one can expect Backbones of smaller cardinality compared to the original solution. We also plan to include other coverage metrics (e.g., area coverage) within the selection process, and load balancing for members of the backbone (i.e., nodes should alternate roles as backbone members in order to prolong the network lifetime).

9. References

Akyildiz, I. F., Weilian, S., Sankarasubramaniam, Y. & Cayirci, E. (2002). A survey on sensor networks, *IEEE Communications Magazine* 40(8): 102–114.

Akyildiz, I., Su, W., Sankarasubramaniam, Y. & Cayirci, E. (2002). A survey on sensor networks, *IEEE Communications Magazine* 40(8): 102–114.

Alzoubi, K. M., Wan, P.-J. & Frieder, O. (2002). Distributed heuristics for connected dominating sets in wireless ad hoc networks, *Journal of Communications and Networks* 4(1): 22–29.

Bandyopadhyay, S. & Coyle, E. (2003). An energy efficient hierarchical clustering algorithm for wireless sensor networks, *IEEE INFOCOM*, pp. 1713–1723.

Bao, L. & Garcia-Luna-Aceves, J. J. (2003a). Topology management in ad hoc networks, *MobiHoc*, ACM Press, New York, NY, USA, pp. 129–140.

Bao, L. & Garcia-Luna-Aceves, J. J. (2003b). Topology management in ad hoc networks, *MobiHoc '03: Proceedings of the 4th ACM international symposium on mobile ad hoc networking & computing*, ACM Press, pp. 129–140.

Basu, P. & Redi, J. (2004). Movement Control Algorithms for Realization of fault Tolerant Ad Hoc Robot Networks, *In IEEE Network* 18: 36–44.

Chen, B., Jamieson, K., Balakrishnan, H. & Morris, R. (2001). Span: An energy-efficient coordination algorithm for topology maintenance in ad hoc wireless networks, *MobiCom '01: Proceedings of the 7th annual international conference on mobile computing and networking*, ACM Press, pp. 85–96.

Chen, Y. P. & Liestman, A. L. (2002). Approximating minimum size weakly-connected dominating sets for clustering mobile ad hoc networks, *MobiHoc*, ACM Press, New York, NY, USA, pp. 165–172.

Chen, Y. P., Liestman, A. L. & Liu, J. (2004). *Ad Hoc and Sensor Networks*, Nova Science Publisher, chapter Clustering algorithms for ad hoc wireless networks.

Clark, B., Colbourn, C. & Johnson, D. (1990). Unit disk graphs, *Discrete Math* 86: 165–177.

Dai, F. & Wu, J. (2005). On constructing k-connected k-dominating set in wireless networks., *IPDPS*, IEEE Computer Society, Denver, CA, USA.

Das, B. & Bharghavan, V. (1997). Routing in Ad-Hoc Networks Using Minimum Connected Dominating Sets, *ICC* pp. 376–380.

Garey, M. & Johnson, D. (1978a). *Computers and Intractability: A Guide to NP-Completeness*, Freeman, San Francisco, California.

Garey, M. R. & Johnson, D. S. (1978b). *Computers and Intractability*, Freeman, San Francisco.

Haynes, T. W., Hedetniemi, S. T. & Slater, P. J. (eds) (1998). *Fundamentals of Domination in Graphs*, Marcel Dekker, Inc.

Heinzelman, W. (2000). *Application-Specific Protocol Architectures for Wireless Networks*, PhD thesis, Massachusetts Institute of Technology.

Henning, M. A., Oellermann, O. R. & Swart, H. C. (1996). The diversity of domination, *Discrete Mathematics* 161(1-3): 161–173.

Joshi, D., Radhakrishnan, S. & Narayanan, C. (1993). A fast algorithm for generalized network location problems, *Proceedings of the ACM/SIGAPP symposium on applied computing*, pp. 701–8.

Li, X.-Y. (2003). *Ad Hoc Networking*, IEEE Press, chapter Topology Control in Wireless Ad Hoc Networks.

Margi, C. B., Petkov, V., Obraczka, K. & Manduchi, R. (2006). Characterizing energy consumption in a visual sensor network testbed, *in Proceedings of the 2nd IEEE TridentCom* .

Paruchuri, V., Durresi, A., Durresi, M. & Barolli, L. (2005). Routing through backbone structures in sensor networks., *ICPADS*, Vol. 2, Fuduoka, Japani, pp. 397–401.

Raghunathan, V., Schurgers, C., Park, S. & Srivastava, M. B. (2002). Energy aware wireless microsensor networks, *IEEE Signal Processing Magazine*, 19(2): 40–50.

Spohn, M. A. & Garcia-Luna-Aceves, J. J. (2006). Bounded-distance multi-clusterhead formation in wireless ad hoc networks, *Ad Hoc Networks (Elsevier)* . http://dx.doi.org/10.1016/j.adhoc.2006.01.005.

Tsuchiya, P. F. (1988). The landmark hierarchy: a new hierarchy for routing in very large networks, *SIGCOMM '88: Symposium proceedings on communications architectures and protocols*, ACM Press, pp. 35–42.

Wu, J. & Dai, F. (2003). Broadcasting in ad hoc networks based on self-pruning, *IEEE INFOCOM* pp. 2240–2250.

Younis, O. & Fahmy, S. (2004). Heed: A hybrid, energy-efficient, distributed clustering approach for ad hoc sensor networks, *Transactions on Mobile Computing* 3(4): 366–379.

11

Graphs for Ontology, Law and Policy

Pierre Mazzega[1,2*], Romain Boulet[3] and Thérèse Libourel[3]

1UMR GET, IRD, CNRS, Université de Toulouse III
2International Joint Laboratory OCE, IRD – Universidade de Brasilia
3UMR ESPACE-DEV, IRD, Université de Montpellier II
[1,3]*France*
[2]*Brazil*

1. Introduction

Since the post-war decades, Public Policies are required increasingly as a preferred tool to promote collective action. Today these public policies are developed through sophisticated participatory schemes involving a variety of actors, public or private. Indeed in the current context of globalization though the State and its administration are key actors (Henry, 2004), their influence fades gradually into a more diffuse institutional environment (Oström, 2005) involving a multitude of other actors (Hassenteufel, 2008). For example at the two ends of the spectrum of governance, we find on one hand the increasing involvement of supranational entities, on the other hand the involvement of nongovernmental organizations.

Therefore, the stated aims of their implementation and the means to achieve them are the result of negotiations and of - at least partial - consensus with stakeholders. In representative governments especially, the trend is growing imposing accountability as regard to the implementation and effects of these policy devices. Following the same trend it becomes mandatory to lean projects of public policy on impact analyses (André et al., 2004; Bourcier et al., 2012) themselves regulated by legal provisions (e.g. European Commission, 2004). These changes induce or reinforce specific phenomena of interest here: a) the systemic nature of law and public policy (and "public action", an expression that might have a more explicit connotation from a dynamic point of view); b) the gradual - yet limited - disappearance of a singular authority (as could be the State) having the monopolistic power necessary for the shift in or the centralized decision-making; c) the emerging nature of the direct or indirect (desirable or undesirable, internalized or externalized, etc.) effects induced by the policy and law.

These introductory remarks that are part of an intellectual position rather than of a due demonstration, lead us to propose a dialogue between the analysis of public policy and the analysis of complex systems. We take the risk of turning to the analysis of law and public policy as complex systems whose implementation mobilizes law, economics, sociology, policy and administrative sciences, and which *understanding* relies on these disciplines but

---

*Corresponding Author

also, in its foundations, on ontology or mathematics (though we are aware that the scale of this ambition is balanced by the modesty of the results presented here).

The modeling of law and public policy is an emerging field of research aiming in particular at understanding their "functioning", at simulating *ex ante* their potential impacts, at finding ways of achieving a desirable goal (back casting), at producing tools for decision support, at allowing the participative drafting of legal texts or policy, especially by using ICT (Information and Communication Technology) or approaches developed for the analysis and structuring of social networks. Two objectives are achieved in this chapter: a) cover part of the literature of recent years in this field; b) illustrate our point with original results drawn mainly from the analysis of law and policy in the wide field of environment and sustainable development.

In Section 2 we present some basic notions in graph theory and network analysis routinely used in this article (and in other cited articles). Our choice is to limit ourselves to a very small number of tools and concepts from graph theory while providing more opportunities for application in law or political science. In Section 3, an ontology-inspired approach is used to represent the law and public policy (and the associated knowledge) in the form of graphs whose analyses are powerful to discover and interpret cognitive and operational structures in these systems. We hope that the dividend of the example will double as our analysis focuses on the representation of current research conducted on the application of ICT in e-governance and modeling of public policies. In Section 4, we present various types of graphs induced by the law and whose analysis leads to propose a quantitative approach to some aspects of legal complexity. In particular the analysis of environmental treaties of international law adds a new graph structure to those previously identified in the analysis of networks of references between laws.

We proceed similarly in Section 5 for public policies considered in terms of organized action. It is in this case a policy of management of water resources implemented in France at the scale of a hydrological basin (large water cycle) where many uses are competing in particular in period of low water. Some perspective for expanding the areas of graph theory that will have a great interest in the modeling of Law and policy are briefly mentioned in Section 6 with concluding remarks.

2. Graphs for the analysis of complex systems

From a general perspective graph theory is today a cornerstone of the study of complex systems. In this chapter we use a few fundamental concepts, measures and estimation methods developed in graph theory and network analysis, that provide us with privileged ways to conceptualize, formalize, model and visualize structural properties of legal and policy systems and to identify their non trivial properties. Both the use and development of these concepts and methods have proven to be fruitful in the study of large interaction networks in sociology or in biological sciences, where emerging structures like small-worlds or scale-free structures are recurrently exhibited. In return, we hypothesize that the contemporary legal and policy "fabrics" pose original problems that may trigger new developments in graph theory and network analysis. In order to show their relevance for the study of legal and policy systems in the next sections, we restrict our "toolbox" to the following limited set of concepts and estimators (the analysis methods and algorithms will be indicated through the cited references only).

2.1 Concepts

A graph is a mathematical object consisting of two sets: a set of vertices (or nodes) and a set of edges (or links), an edge linking two vertices. This mathematical structure of graph and its related analysis tools are used to model real phenomena of interaction between objects that are networks (see e.g. Brandes and Erlebach, 2005). We now introduce basic vocabulary of graph theory that will be used in this chapter. An induced *sub-graph* H of a graph G is a graph whose set of vertices is included in the set of vertices of G and there is an edge between two vertices of H if and only if there is an edge between these two vertices in G.

An undirected graph is *connected* if there is a path between any two vertices. A *connected component* of a graph is a sub-graph of maximum size (in particular a connected graph has one and only one connected component: itself). A directed graph is *weakly connected* if the underlying undirected graph is connected. It is *strongly connected* if there is a directed path (in each direction) from each vertex to any other vertex. A *strongly connected component* is a maximal strongly connected sub-graph.

A vertex v is a *neighbor* of a vertex u if there is an edge between u and v. The *neighborhood* of a vertex u is the sub-graph induced by the set of its neighbors and the *degree* of u is the number of its neighbors. A *complete graph* is a graph where each pair of vertices is linked by an edge. A *clique* in a graph is a complete sub-graph. Several matrices can be associated to a graph such as the adjacency matrix A whose (i,j)-element is 1 if vertex i is linked to vertex j and is 0 otherwise.

2.2 Measures

Understanding the architecture of a network first passes through a structural analysis. Networks are of many different kinds - simple, directed, weighted, labeled, etc. - which requires the development of specific indices to measure their properties. Some indices measure basic characteristics of the network such as the number of vertices or nodes n, the number of edges or links m, the density d of the network (defined by the ratio between the actual number of links and the total possible number of links) and the mean degree.

The indices measuring the global connectivity of the network are built on the notion of shortest paths in the network. A *shortest path* between two vertices is a path (a sequence of consecutive edges) of minimal length. Global connectivity is then evaluated by the *mean of the shortest paths lengths l*, the *characteristic path length L* (median of the means of shortest paths; Watts, 2003) or the *diameter D* (the length of the longest shortest path in the graph). Two indices, called clustering coefficients, measure the local connectivity. The *first clustering coefficient $C1$* is defined by the average density of the neighborhood of a vertex. *The second clustering coefficient $C2$* is the ratio between the number of triangles in the network and the number of connected triples in this network which can also be seen as the probability to have a link between two vertices linked to a common vertex.

In order to evaluate the position of a vertex in a graph, three notions of centrality have been defined (Freeman, 1979). The degree centrality C_D measured by the degree of a vertex, the betweenness centrality C_B related to the number of shortest paths going through a vertex and the closeness centrality C_C of a vertex based on the mean distance to the other vertices. Then the centralization of a network is the normalized sum of the differences between the vertex with the highest centrality and the centralities of the other vertices in the graph. A zero (resp. unit) value of the network centralization corresponds to a non centralized (resp. most centralized) network. The *degree distribution* is the probability distribution of the vertex degrees; for each integer k it gives the probability to have a vertex of degree k. Other indices may be introduced dealing more particularly with the impact on the network structure of weights or the presence of different kinds of links (Boulet, 2011; Boulet et al., 2011b).

2.3 Finding communities

An essential step in network analysis is the research of communities. In a social network, some people can be gathered into communities. Following this example of community we can extend this notion to networks of different nature (networks of legal texts, lexical networks, etc.) analogously defined as groups of vertices more densely connected to each other than the rest of the graph. Detecting communities in networks is a very active field of research in network analysis and we shall consider only a few of the existing algorithms.

Several algorithms have been developed relying on various methods. The walktrap algorithm (Pons and Latapy, 2005) is based on random walks on graphs with the underlying idea that a random walk would be trapped into a community. The fast greedy algorithm (Clauset et al., 2004) aims at greedily maximize the modularity (a criterion assessing the quality of a partitioning) of the resulting partitioning. Finally spectral methods (von Luxburg, 2007) are based on eigenvalue decomposition of matrices associated to a graph which embeds the graph in an Euclidean space on which we can use statistical clustering.

A special community is worth mentioning here, the *rich-club* (Zhou and Mondragon, 2004). A rich-club is formed when the vertices with highest degree (the rich vertices) are highly interconnected and form a very dense group (a club). They therefore constitute a central and influent community in the network.

2.4 Types of graphs

An *Erdös-Rényi random graph* (Erdös and Rényi, 1959) is a graph where an edge is put between two vertices with a uniform probability p. It is known (Watts, 2003) that such a graph have a tight global connectivity (the randomness of edges creates shortcuts) and low clustering coefficients (the local density looks like the overall density). Then the indices presented earlier allow distinguishing several classes of graphs. A *small world* network (Watts, 2003) is a network with a tight global connectivity (nearer than that of a random graph) and a high local connectivity (much larger than that of a random graph). A *concentrated world* (Boulet et al., 2011a) is a dense graph with some highly interconnected vertices having the highest centrality measures. A *scale-free network* is a graph with a power-law degree distribution. Therefore it has a lot of vertices with a low degree and few vertices with a high degree (sometimes called hubs). It contrasts with Erdös-Rényi random graphs, the degree distribution of which is centered to a mean value and follows a Poisson distribution. A directed graph is a graph with directed edges. A special kind of directed graph is a tournament in which there exists an edge between any two vertices (it can be seen as a complete directed graph).

3. Ontology as a network

The title of this section calls for comments. First, even if we exclude *a priori* from the scope of our discussion the field of philosophy, the term "ontology" has no unambiguous acceptance. By the 1980s, the field of computer engineering and knowledge representation takes the term to define a computational model for performing automated reasoning. In 1995, T. Gruber says that "an ontology is a formal, explicit specification of a shared conceptualization" (Grüber, 1995). In 2009, he refines his remarks by stating that "an ontology defines (specifies) the concepts, relationships, and other distinctions that are relevant for modeling a domain. The specification takes the form of the definitions of representational vocabulary (classes, relations, and so forth), which provide meanings for the vocabulary and formal constraints on its coherent use" (Grüber, 2009). Finally, the pragmatic issues converge in the statement that "an ontology is a tool and product of engineering and thereby defined by use ». The key issues opened by these proposals relate to how to make the best choice of language, and how to set objectives of use of these ontologies.

3.1 Languages and objectives for ontology

In terms of languages, it is interesting to quickly scan the history. Originally, we find the semantic networks (Quillian, 1968) which are doubly labeled directed graphs, the vertices being labeled by concepts, the arcs being labeled according to relators between concepts. Then follow, the most significant proposals emanating from Minsky, Sowa, Brachman and Levesque. Minsky (1975) proposes an approach based on the notion of a grain of knowledge or frame and specifies that « here is the essence of the theory: when one encounters a new situation (or makes a substantial change in one's view of the present problem) one selects from memory a structure called a frame. A frame is a data-structure for representing a stereotyped situation. Attached to each frame are several kinds of information. We can think of a frame as a network of nodes and relations. The "top levels" of a frame are fixed, and represent things that are always true about the supposed situation. The lower levels have many terminals–"slots" that must be filled by specific instances or data».

Sowa (1984) proposes the conceptual graph formalism. Those are bipartite graphs with two types of nodes: concept-nodes and relation-nodes. Knowledge representation can be defined in a specific area, this being reflected in what is referred to as a support and includes two lattices: the concepts and relationships pertaining to the chosen field. The reasoning is based on semantics in first order logic, but a more original view suggests compensating the mechanisms of logical inference by graph operations in particular the homomorphism of graphs (called projection; Chein and Mugnier, 2009). Brachman (1979), Brachman and Levesque (1985, 2004) propose extensions to the language of frames and semantic networks from semantic-based description logics. They introduce the notions of concept, role and individual and rely on an inference mechanism based in particular on subsumption.

These theoretical propositions are the heart of various studies conducted to date. They are accompanied by proposals for operational developments, the most iconic of them being probably KIF (Knowledge Interchange Format) on the initiative of the consortium DARPA Knowledge Sharing Effort (Genesereth, 1992), the Ontolingua project (Gruber, 1992) until the arrival of the Semantic Web following the original article by Berners-Lee et al. (2001), which paves the way for many of the proposals in terms of protocols, languages, standards (XML Extensible Markup Language, Resource Description Framework RDF, RDF Schema, OWL Web Ontology Language) and tools. In the field of legal studies these researches are now flourishing (Casanovas et al., 2007; Sartor et al., 2011).

Let us return to the question of the use and purpose of ontology design in the context of knowledge of public policies. It is now commonly accepted (Guarino, 1998; Gangemi et al., 2001) that ontologies can be stratified by level of generality or abstraction (Figure 1): a) top-level ontologies describe very general concepts like space, time, material, objects, events, actions, etc., which are independent of one problem or particular application domain. They specify very general abstract knowledge whose content depends on the degree of formalization; b) domain ontologies and problem-solving ontologies or task ontologies that describe, respectively, the knowledge related to a generic domain (like medicine or aviation) or generic activities (as water management) by a specialization of the concepts presented in the top-level ontologies; c) application ontologies describe concepts depending both on a field and on a particular task. They are specializations of the two previous types of ontologies.

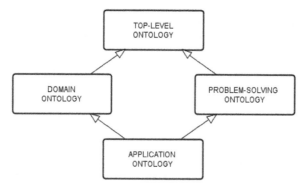

Fig. 1. Scheme of the stratification of ontologies by levels of generality and abstraction.

Overall, it appears that the graphs are essential in the development of knowledge representations and ontologies, and that the synergy between different approaches is important (Farquhar et al., 1995). The example we now develop tackles with the representation of a field of knowledge. The approach taken may not be consistent with the *doxa* but illustrates the types of exploration that allows the use of graphs in the field.

3.2 Classification and phenomenological graphs

Let us return to some fundaments. We here call "classification graph" a graph built on the relationship of subsumption ("is a" relationship) that binds an instance to its class membership. According to an Aristotelian-type classificatory approach, a concept (e.g. an agent of the administration) may be both a special case - one instance - to a larger class (the agents) and a (sub-) class for other more specific instances (e.g. the prefect). The support structure of these univocal classifications is of course the tree-graph that we propose to call "class graph". According to its purpose, this structure is effective and unambiguous.

But by destroying any flexibility of interpretation, this normativeness may be undesirable. Let us consider an example which is specific to Law. The judge is asked to reinterpret terms or concepts in a manner more consistent with the acceptance that tends to prevail in a constantly evolving social context. Thus the "family" of the Civil Code of 1804 is no more precisely that *family* of a society that tends to recognize the single-parent family, same-sex parents or surrogate mothers. Sometimes it follows a reassessment of existing case law, the judge taking a unique position that other judges will adopt later. For schematic that can be our example (indeed, the departure from precedent is less about the interpretation of terms than about the reinterpretation of a point of law which, presumably, involves an entire ontological environment), a sociologically-oriented ontology should be able to allow, in the organization of its classes and of their instances, the occurrence of multiple relations of subsumption (one instance being subsumed to several classes), in a way more conform to uses than in line with normativeness of any *a priori* model (should it be dominant).

Incidentally, we move from the trees (and forests) to the wider world of all graphs. These graphs then capture the structure of a conceptualization of a system that relies on a variety of types of objects and typed relations, a "model" or "ontology". In this chapter we propose to call these structures "phenomenological graphs" – or "*pheno-graph*" for short - because they capture the observable structure (whatever the means of observation) of relationships between entities whose existences are attested.

There are various ways to build such a model, based on pragmatic approaches in the fields of law or public policy. One of them is of particular interest in this section. It is based on the terminological analysis of large corpus of texts (legal texts, texts describing public policies, regulations, etc.). The analysis of terminology and text mining can also be based on a representation of the knowledge domain in the form of ontology, the text analysis revealing more or less reinforced relationships between terms or concepts (or even invalidating the relevance of some ontological relations in this context or by inducing new links).

Formally, this type of analysis consists of two levels say a tree-like classification for supervision and the texts revealing a more general network of entities linked by heterogeneous relationships (such double structure is described in Mazzega et al., 2011). Let us now work an example of a double structure – ontological and phenomenological - in the field of research mapping.

3.3 Illustration: Research in e-governance & policy modeling

The CROSSROAD European project produced a public report on the state of the art on the field of ICT research in electronic governance and policy modeling (CROSSROAD, 2010) from which we here extract and analyze the data structure. This research field is divided in five domains divided in areas and sub-areas as follows: 1) Open government information and intelligence for transparency (4 areas; 18 sub-areas; *associated color:* orange); 2) Social Computing, citizen engagement and inclusion (4 areas; 14 sub-areas; green); 3) Policy modeling (4 areas; 18 sub-areas; yellow); 4) Identity management and trust in governance (4 areas; 18 sub-areas; blue); 5) Future internet for collaborative governance (5 areas; 25 sub-areas; purple). In this list we attribute a color to each research domain and to its afferent areas and sub-areas for the purpose of graph visualization.

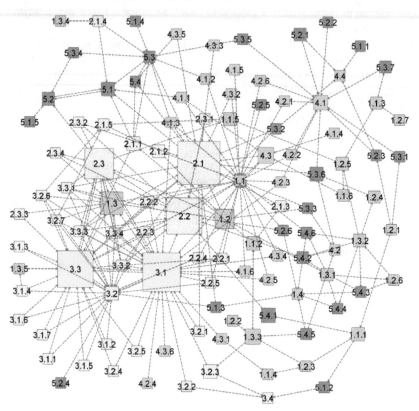

Fig. 2. A representation of the *pheno-ICT graph* with the size of each vertex being linearly related to its betweenness degree (see text). Vertices of the same color belong to the same research domain.

The full mapping of this field of research shows the type of double structure mentioned above, say: a) a tree structure - that we call *onto-ICT graph* (not represented here) - induced by the organization of this field of research in the above mentioned 5 domains (see below), 21 areas and 93 sub-areas (grouped under the term "research topics") that are the 119 vertices of the graph, with 114 edges; b) a network structure – which principal connected component we call *pheno-ICT graph* (111 vertices, 245 edges) - in which topics (research domains, areas or sub-areas) are linked by pairs when the specific researches they cover are explicitly related in a noticeable amount of scientific publications or when these publications refer to specific terminologies preferentially associated with these topics.

In Figure 2 we give a representation of the *pheno-ICT graph* that enhances the relative magnitudes of betweenness centrality associated to the vertices. The correspondence between the labels of the vertex ("x" type label for domain,"x.x" for area, "x.x.x" for sub-area) and research field they cover can be found in CROSSROAD (2010). Let us now imagine some simple scenario (about research policy for policy research). Should we organize some meeting in ICT research in electronic governance and policy modeling, what should be the title of the sessions? A convenient choice might be to choose those topics with the highest degrees in the *pheno-ICT graph*. If now some (financial or human) resources are available to strengthen research where should it be primarily allocated? Such a choice of course depend on the criteria for decision making but a strategy could be to support those topics that present a high betweenness centrality, because they will reinforce the synergy between various research communities and strengthen the whole research system.

Label	Name of the research topic	Betweenness Centrality Index (rank)	Normalized Degree (rank)
2.1	Social Computing	1.00 (01)	0.64 (05)
3.1	Policy Analysis (in policy modeling)	0.86 (02)	1.00 (01)
3.3	Visualization (in policy modeling)	0.80 (03)	0.75 (03)
2.2	Citizen Engagement	0.78 (04)	0.46 (06)
2.3	Public Opinion Mining & Sentiment Analysis	0.60 (05)	0.39 (07)
1.3	Visual Analytics (in open government information)	0.36 (06)	0.39 (07)
1.2	Linked Data (in open government information)	0.26 (07)	0.14 (13)
4.3	Trust (in governance)	0.16 (08)	0.11 (17)
1.3.3	Analytical Reasoning (in visual analytics)	0.14 (09)	0.07 (22)
4.1	Identity Management	0.12 (10)	0.36 (09)
5.4.1	Web Accessibility (in human / computer interactions)	0.11 (11)	0.07 (22)
5.3.6	Public Service Aggregation, Mash-ups & Orchestration	0.10 (12)	(>25)

Table 1. Ranking the first twelve research domains, areas or sub-areas of ICT research in electronic governance and policy modeling by decreasing (normalized) betweenness centrality index (as extracted from the pheno-ICT graph) considering only entering links. The normalized degree (and rank) of the research topics are also given for comparison.

In Table 1 we give the first twelve research topics ranked by decreasing (normalized) betweenness centrality index. The normalized degree (and corresponding rank) of the research topics are also given for comparison. It appears that though they do not have a high rank of connectivity, several topics have a high rank in terms of betweenness centrality. This is the case for the first topic "social computing" (betweenness centrality rank=1; connectivity rank=5) that is about the application of web social software in public-sector activities and the support to social interaction and collaboration (and with the associates terminology: social networking, content syndication in government portals, blogging and micro-blogging, collaborative writing tools, feedback, rating and reputation systems – CORSSROAD, 2010). It is interesting to notice also that several topics among the most connected with the other topics are not in the short list of high betweenness centralities: area 1.1 "open & transparent information management" with degree 0.89 (2d rank), area 3.2 "(policy) modeling and simulation " (degree 0.75 – 3rd rank), area 5.3 "multi-channel access and delivery of next generation of public services " (degree 0.32 – 10th rank), areas 1.4 "findings (in open government information and intelligence for transparency)" and 5.1 "cloud computers (for collaborative governance)"(degree 0.21, 11th rank).

	Index	Non-oriented *pheno-ICT* Graph	Average Indices of *Erdös-Rényi Graphs*
General Characteristics	n	111	109.56
	m	239	239.08
	d	0.039	0.040
	k	4.31	4.36
Global Connectivity	l	3.25	3.34
	L	3.15	3.24
	D	8	6.92
Local Connectivity	C1	0.34	0.092
	C2	0.12	0.038
Network Centralization	C_D	0.24	0.055
	C_B	0.11	0.037
	C_C	0.27	0.15

Table 2. Small world indices calculated on the non-oriented *pheno-ICT* graph and for comparison on a set of 1000 *Erdös-Rényi* random graphs with the same size and order. n is the number of vertices, m the number of edges, d the density, k the average degree, l the length of the average shortest path, L the characteristic length, D the diameter, C1 the first clustering coefficient, C2 the second clustering coefficient, C_D the degree centrality, C_B the betweenness centrality and C_B the closeness centrality (see Sec.2).

On one hand these simple findings can be useful for designing a research policy in this field, as briefly suggested by our quite trivial scenarios or by many others related to real life situations. On the other hand it is interesting to see if the pheno-ICT network belongs to a known class of graphs. We consider the undirected graph (some geodesics do not exist on the directed graph) for which we assess the general indices as well as the global and local connectivity (Table 2). The random model used for comparison is the G(n,p) model of *Erdös-Rényi* (See Sec.2) where there is an edge between two vertices with a probability p equal to the density of the graph. The examination of these connectivity indices leads us to conclude that this network is a *small world*. Indeed, it fills the following two conditions: a) tighter global connectivity similar to that of random graphs; b) strong local connectivity, much higher than that of a random graph. So we are dealing with a type of graph which shows the structure of many social networks the study of which is enriched with other concepts opened to interpretation, but that we will not discuss in this chapter.

We also observe that the (non-oriented) *pheno-ICT* network exhibits high values of the betweenness centrality (but not of the closeness centrality) when compared to random graphs. We interpret this indicator as showing that the work developed in many topics (domains, areas or sub-areas) are routinely called upon to involve or interest not only directly related topics but also other neighboring topics albeit slightly more distant (on the network). In a sense, this feature reflects the strong identity and overall coherence of research in the field of ICT research in electronic governance and policy modeling.

4. Law-induced networks

According to the classic positivist theory of law (knowing that we do not consider here the Common Law), legal norms must obey a hierarchical organization implying that the norms of lower degrees should not be in conflict with the norms of higher degree (Kelsen, 1960). At the top of the pyramid of norms is the Constitution, more or less expanded with a "block of constitutionality", followed by the organic and other laws, decrees, etc. With the rise of Community, European and International law, also with the proliferation of sources of law, this hierarchical organization is deeply upside. The doctrine echoes these on-going transformations (eg. Ost & van de Kerchove, 2002; Delmas-Marty, 2007) that are also subject of much debate in the democratic representative bodies.

For example in France, the positions of the Court of Cassation and of the Council of State have been for nearly 15 years in opposition as to the primacy of European law over national law, until the adoption of a common position recognizing it during a reversal of precedent of the State Council (see Bécane et al., 2010). Of course this rule is accompanied by measures not to amputate the representative bodies of their ability to legislate or to exercise control over norms produced outside their assemblies, the national legislature thus acquiring a role in the European legislative process (Article 88-4 of the Maastricht Treaty).

So we perceive that under the combined effects of globalization and the empowerment of communities and organizations (e.g. NGOs) network structures appear in all domains of Law and policy. Different kinds of networks: institutional networks (the link being instantiated by roles, powers, information flows, etc.), networks associated to the process of decision-making (e.g. EU), networks and flows in law-making, etc. But outside of these role and power games the analysis of which is subtle, another phenomenon reflecting the overlapping of legal systems can be easily observed and studied. Although its range is relatively minor (it is of interest for some operational services of law broadcasting like LEGIFRANCE www.legifrance.gouv.fr/, EURLEX eur-lex.europa.eu/en/index.htm, etc. and also for codification), its analysis allows starting the development of measures of certain manifestations of "legal complexity": this phenomenon is the emergence of networks of citations between legal texts (Bourcier and Mazzega, 2007a; Bommarito and Katz, 2010). The citations between legal texts results from a self-organizing process, no top authority being in charge to manage the constant flow of legislative regeneration, neither at the national nor regional or global level.

Based on previous work, we will support the next two assumptions that do not deny the concerns of the science involved in the analysis of complex systems, namely: a) that the types of structures found in the network of citations depend on both the size of the object in question and the scale of resolution for their analysis; b) that the "canonical" representation of a field of knowledge - in this case the legal knowledge - deserves to be coupled with a representation based on the analysis of emerging phenomenological graphs. We will then give further arguments in favor of these two hypotheses, based on an empirical analysis of the bipartite network of environmental treaties of international law.

4.1 Scale matters in law

Let us start with small scales. The article is classically considered as the smallest entity that deployed in a broader frame – law, code - should have a legal content making sense of its own. Yet we have found up to 4 levels of subdivision in a significant number of articles in the environmental code, and we have estimated their frequency distribution and length of text statistics (Bourcier and Mazzega, 2007b). The association of a graph of citations to a corpus relies on the choice of the resolution for analysis: should we keep all the subdivisions or stop at the granularity of articles (with 1266 of them in the environmental code, legislative part)? This choice is in fact dictated by the objectives of the thematic analysis. However we understand intuitively that the absorption of certain entities (e.g., subdivisions of articles) in the entity of which they are parties (e.g. article) increases the density of the network by aggregating the parts in the local whole and assigning all edges to it.

On the other side of the spectrum of sizes, the choice of a definition of the analyzed object is just as important. The environmental code has in a sense a unity conquered by the legislature (especially with regard to the rural code) but partly arbitrary since it cites (or is quoted by) 35 other codes, several EU Directives, international treaties etc. Considering at a given time all of the existing law in the world and its ramifications (to which corresponds a global legal-graph), being interested only in the French environmental code, or in all of the codified French law (which associated graph we call hyper-code), is equivalent to extract an induced sub-graph and analyze it separately. This sub-graph can be seen as associated to an ontology-derived community which does not have the emerging character of the phenomenology-derived communities we will consider below. Indeed the ontology-derived communities are deliberately constructed by the legislature, or in this case, coder (two "actors" that deserve being considered themselves as communities of actors).

The use of approaches presented in Section 3 also allows identifying vertices with remarkable properties: for example, articles (resp. Codes) of high degree centrality or high betweenness centrality in the environmental code (resp. hyper-code graph; Mazzega et al., 2009; Boulet et al., 2011a). Thus, as in many complex systems the choice of the analysis resolution and of the size of the object changes the characteristics and properties of the associated graphs, but also the graph type and hence the paradigm of interpretation.

4.2 Alternative representation of legal knowledge

The matter of a code - as the Environmental Code - is usually organized into divisions - books, titles, chapters, sections, articles - that form a tree structure (table of content). However, another organization is behind it, which brings together the divisions with a high density of inter-citations (the main connected graph component associated with the legislative part of the Environmental Code has 980 vertices and 2186 edges). The use of multiple algorithms for partitioning the graph associated to a corpus, allows identifying stable (say found by all algorithms based on different criteria for partitioning - cf. Sec. 2.2) "hidden" communities of its divisions - including articles - which are semantic units pertinent for the interpretation of the law.

Two points are worth noting: a) the existence of these phenomenological communities is known neither to the legislature nor to the lawyer, b) some of these emerging communities do not coincide with the ontological divisions, necessary in the planned organization of the code (Boulet et al., 2010). Conversely some ontological divisions are not in any stable community, such as for example the provisions relative to New Caledonia or Antarctic (a result that we keep to interpret). In other words, the codified substance could be distributed in a different cognitive representation and determined according to objective criteria (reproducible between codes - possibly from different countries).

Similar results are obtained when analyzing the hyper-code graph of citations among all codes of the French legal system produced using the same codification methodology and whose vertices are the codes (52 to date). Thus we have mainly highlighted (Boulet et al., 2011a,b): a) a rich-club (see Sec. 2.3; Colizza et al., 2006) including the 10 most central and influential codes of the French legal system; b) a stable community of 12 codes governing the legal areas linked to social activities, their regulation and security; c) a second stable community of 11 codes governing the legal areas related to land management, the territories and their resources. Only the analysis of hyper-code code could lead to identify those communities whose existence should challenge the doctrine, their existence being previously unknown.

4.3 Illustration: A tournament of international treaties

We now consider the state (as of 2010) of ratification of the 42 environmental international treaties by the 196 countries (or entities of equal status, such as the European Community) members of the United Nations (see the Chapter XXVII of the database of international treaties of the United Nations http://treaties.un.org/Pages/ParticipationStatus.aspx). Initially we built a graph with a link between countries A and B if both have ratified at least one common treaty. The links are then weighted by the number of common treaties ratified by the two countries. The partitioning of this graph showed two groups (Boulet et al., 2012): one group of 38 countries with only Canada being not part of continental Europe, and a group bringing together 158 countries.

In a second step, we seek a graph showing a higher discriminatory power of any developed strategies for ratification by member countries. For this we used the following rule: a link is established from country B to country A if B is the first country to ratify a particular treaty after ratification of that treaty by A. A second rule must be established with respect to N countries simultaneously ratifying the same treaty: a) link the N countries, which induces a clique in the graph or b) not link the N countries, which induces a stable graph. In the following analysis we use option (a) as our second rule. The graph (196 vertices, 2832 links) we obtain is shown in Figure 3. The hypothesis that motivates the analysis of this graph is that the sequence of dates of ratification reveals (whether collaborative or competitive) collective political strategies of countries (Louka, 2006).

The five countries most ready to sign the environmental treaties are (ranked by decreasing order of the degree centrality) Norway, Finland, Hungary and at the same level, France and Luxembourg. The countries with the highest betweenness centrality, and thus being somewhat in the stream of the temporal succession of ratifications are (ranked by decreasing order of betweenness centrality) Hungary, France, Belgium, Romania and Norway. These simple results show unambiguously the active role of Hungary, Norway and France about the environment on the international stage.

We find four stable communities with 41, 39, 22 and 10 countries respectively. Most European countries (in a geographical sense) are the second community, alongside the United States, Canada, Japan, New Zealand and Burkina Faso. The contrasting positions of these countries with regard to Europe, in particular on issues related to the climate change and energy policy suggests that this grouping into a single community rather reflects an antagonist-type of interaction. Note that this community also has the largest clique of the graph composed of 18 countries (Germany, Austria, Belgium, European Community, Denmark, Spain, Finland, France, Greece, Ireland, Italy, Japan, Luxembourg, Netherlands, Portugal, United Kingdom of Great Britain and Northern Ireland, Slovakia, Sweden).

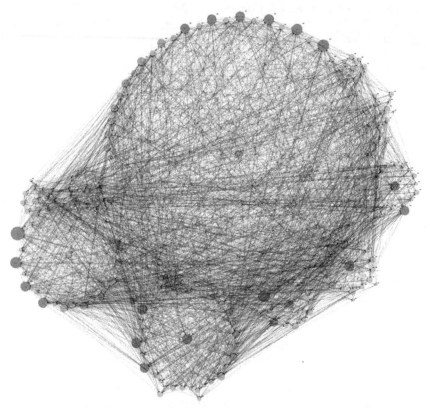

Fig. 3. A representation of the oriented graph associated with the ratifications of the environmental treaties and conventions of International Law. The vertices are 196 countries (or analog legal entities like the European Community) of the world. A link is drawn from B to A if B is the first country having ratified a given treaty just after country A. The larger orange (resp. smaller green) vertices represent countries with the highest (resp. lowest) betweenness centrality (see text).

Among the emerging countries, Brazil, China and India are in the third community (22 countries), but not South Africa (in the 4th community). The Russian Federation does not appear in any stable community (as defined here in Sec. 2.3), as also 83 other countries.

Countries suffering from acute chronic internal conflicts, backed to failing States, are among the countries that have ratified the least environmental treaties, a fact that reveals a simple inspection of their degree centrality. The logic behind the arrangement of the first community, made up of many countries from different geographical, geopolitical or economic regions, is not easy to identify. This probably results from the existence of several smaller sub-communities but whose ratifications are interspersed.

So if these early results are promising (as we think about the other 28 chapters that bring together hundreds of treaties), a more detailed analysis of these data, taking into account the particular dates of ratification themselves and a diachronic knowledge of the major world events, should identify more subtle patterns of international relation strategies as expressed by the evolution of international law (Schneider & Urpelainen, 2012).

5. Graphs for policy analysis

The analysis of citation networks between legal texts foreshadows network analysis for components and links of heterogeneous networks that may be associated with public policy. We have already mentioned that throughout the life cycle of policy (agenda-setting, formulation, decision making, implementation, evaluation; Howlett and Ramesh, 2003) are mobilized a variety of actors (individual or group, institution, etc.) and of resources.

One objective of these models is to provide tools for the *ex-ante* assessment of the potential impacts of these policies. This ambition is very high and we believe should be seen as a horizon to strive for rather than an achievable goal in the near future. Indeed at the individual scale, actors' behavior is unpredictable, their resilience and creativity never limited even if in a highly regulated or binding frame. However, at the collective level, patterns of behavior are observed and can be analyzed using proven methodologies. The analysis of graphs associated with the representation of certain aspects of public policy adds to the available (or developing) tools promoting a better understanding of these instruments of the "public" collective action. Rather than remain at this level of generality we prefer in this section to develop an example who keeps illustrative in general. This is the development of a simulation platform of the impact of new norms for managing water resources at the scale of a hydrological basin.

5.1 A framework for water policy

The water management in France is governed by various legal provisions, including the Law on Water and Aquatic Environments 2006 (LEMA, 2006) which adopts tools to achieve by 2015 the goal of "good" water state as set by the European Water Framework Directive (WFD, 2000). Basically the water management is organized at the scale of large hydrological basins. Southwest of the French metropolitan territory, the Adour-Garonne basin (AGB) covers about 120 000 km², with a rate of precipitation of 600 mm to 2000 mm/year, a potential for runoff of 90 km³/year and large stocks of water in the aquifers. About 7 million people are distributed over 6900 municipalities and 35 cities with more than twenty thousand inhabitants (most such data are provided by the website of the Adour-Garonne Water Agency www.eau-adour-garonne.fr/).

Average annual withdrawals are 2.5 km³, being roughly equally distributed between the three main uses, namely drinking water, industrial uses and agriculture. However, the water availability and uses are very uneven over the year: during the low water period (summer and early autumn) agricultural levies represent 80% of the withdrawals for about 645 000 irrigated hectares. During the heat wave of 2003 the withdrawal for irrigation has nearly doubled compared to average years. On the other hand, the climate change scenarios predict that 2003 will be in terms of rainfall better than the average year of the 2050's (Pagé & Terray, 2011). Consequently, the management of low water is the first priority of the Adour-Garonne Water Agency. The water uses must also be balanced with the preservation of aquatic ecosystems and environmental services, two issues that the EU directive and the LEMA (2006) explicitly raised as priorities.

On the scale of the Adour-Garonne basin, the water policy and its implementation was prepared with the participation of governmental administrations, territorial and other communities, associations, the civil society, etc., over several years, as described in the master water planning and management report (SDAGE AG, 2010) and in the associated implementation program (PDM AG, 2010), documents whose provisions may be legally opposed to other sectoral policies (territorial development, economic development, etc.). More locally, across the river basin management units (sub-basins), the low-water management plans specify the measures to regulate water uses within the environmental, social, economic and ecological constraints associated to periods of low flow.

In the MAELIA project (Multi-Agents for EnvironmentaL norms Impact Assessment) we develop a hybrid model combining a multi-agent system, model equations and GIS to estimate the societal, economic and environmental impacts of new "norms" implementation in the basin. We focus specifically on the new device that will limit water withdrawals for irrigation in many watersheds. To evaluate *ex ante* the potential impacts of this new frame, we represent in the model both the behavior of key actors involved in the water uses or management, but also the environmental (including hydrology) and climate dynamics as well as the main activities (especially agriculture) developed by the actors that may have a direct or indirect effect on the resource. To achieve this goal it is first necessary to identify the key actors, the activities they lead, the resources (material and cognitive) they use or generate, and the social and environmental processes the entire system depends on (Sibertin-Blanc et al., 2011; Thérond et al., 2011). This step was carried out by an interdisciplinary participatory method that results in the production of a conceptual model presenting only the essential elements of the management of low water in the basin.

5.2 Illustration: Basin-scale low water management

The conceptual model is built up on the basis of various UML diagrams (OMG, 2005). Figure 4 shows the main one, namely the actor- resource diagram (see Sibertin-Blanc et al., 2011 for some other diagrams). The vertices are instantiations of the following classes: actors, material resources, cognitive resources (including the norms: legal texts, programs, directives, etc.). The links are of different types (with cardinalities that we discard here): actors' activities, social processes, environmental processes or conceptual relationships between the linked entities. We are interested in the structure of the graph associated to this conceptual model of the low water management in the Adour-Garonne basin.

The graph has a single component of 66 vertices and 87 edges. It is weakly connected, with only one strongly connected group of vertices formed of two material resources (equipment for drinkable water catchment; equipment for raw water catchment and flow like canals, pipes, etc.), two "actors" (remembering that the term « actor » can refer to a group of actors: raw water distributors; clean water distributors) and one cognitive resource (structure of water prices and charges). Functions or operations associated with the vertices with the highest degree of connectivity must be modeled accurately because they directly affect the states of many others. These are in particular (by decreasing degree) the farmers, the (public or private) equipments, socio-professional categories (regarded as consumers of drinking water in municipalities), all types of water tanks, the land sub-parcels.

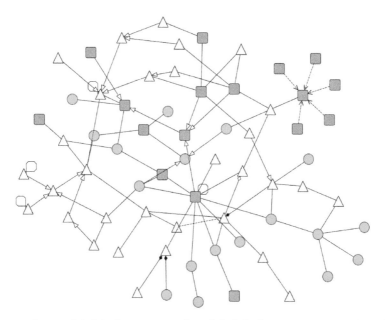

Fig. 4. The graph associated to the conceptual model of the law water management in the Adour-Garonne basin (France). The colors of the vertices indicate their type: actors (orange round rectangles); cognitive resources and norms (green circles); material resources (yellow triangles). The labels, attributes and functions associated to each vertex in the UML model have been removed to clarify the graph (some of them being given in the text).

On the distribution drawn on Figure 5 we see that around ten vertices present a significant betweenness centrality index. These vertices are the following ones: (label=49) Raw water distributors; (09) Manager *of* engineering structure (dams, etc.); (61) Manager of water catchment; (47) Clean water distributors; (57) Water tanks; (07) (Public or private) Equipments; (14) Equipment for water catchment; (17) Equipment for drinkable water catchment; (00) Geo-referenced territorial unit; (01) Plots. From a perspective of how the dynamical model works, the behavior of these entities (including changes in their respective states and comparing them with empirical data) should be given special attention. Indeed their intermediate position a priori designate them as being likely to exhibit abnormal behavior in the case all processes and activities modeled are not adequately represented or adjusted against each other.

A look at the global and local connectivity indices estimated for this graph does not allow us to characterize it as a small-world network. Although the overall connectivity is tightened (L= 4.14, D = 8) and close to that of random graph (L = 4.58, D = 11.32), the local connectivity is low (C1 = 0.41 C2 = 0027): the first clustering coefficient is artificially high due to the large number of vertices of degree 1, the second clustering coefficient is low. Moreover, the graph contains only two triangles.

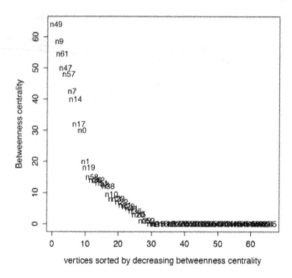

Fig. 5. Distribution of vertices of the low water management graph (Fig.3) in descending order of their degree of betweenness centrality.

All these results concern the static structure of the conceptual model of low water management. From the conceptual model is then derived a dynamic model where all the social and environmental processes (represented by computer codes) change over time the state variables (or attribute values) of the actors and resources. The analysis of the dynamic properties of this type of model must use a range of other tools (Mesbahi and Egerstedt, 2010), including statistics we will not address here, but whose scope covers any multi-agent system, being dedicated to the modeling of public policies and of their potential impacts or to any other complex system dynamics.

6. Conclusion

As was briefly shown in this chapter network analysis is useful for many purposes in Law and Policy: to identify emerging cognitive sets or patterns (groups of concepts); to make a functional mapping of the components of a complex system and of their interaction; to carry out all stages of modeling (from the definition of a referential representation to the setting up of a conceptual model and its implementation in a simulation platform for impact assessment) ; to analyze social dynamics (like strategies, coalition dynamics, etc.).

Many other concepts can be requested from graph theory (Jungnickel, 1999) to highlight the hidden properties of legal corpuses or public policies. For example we mentioned in Sec.4 the impact of the aggregation of nodes (corresponding to different choices of resolution) in a root vertex on the density and other measures associated to a network. A similar transformation occurs when passing from a given graph to the graph of its stable communities. Graph theory can provide fundamental knowledge on the incidence of these operations on the main measures associated with the studied networks, which would constitute a major theoretical and methodological contribution to the analysis of complex legal and policy systems.

As soon as graphs are involved in formal representations, their comparison provides a methodology for comparing the complex legal and policy systems they capture. Such issues appear especially in ontology alignment, in the identification and analysis of functional similarity of actors' positions in governance schemes, in the comparison of policies or legal systems between different sectors or countries, etc. Techniques of graph mining (Cox and Holder, 2007) would be particular useful in conjunction to text mining, a huge amount of data being accessible today via institutional public web sites and data base systems (as illustrated in this chapter with LEGIFRANCE, EURLEX or the UNO data basis on the multilateral treaties).

Another exciting area of research is to provide a graph-related equivalent to concepts developed and tested in the social sciences, especially law and political sciences. Different schools of thought have developed conceptualizations of central concepts such as power or role (among many others), which could be translated in terms of graphs and their properties and thus be enriched by the theoretical contributions of mathematical research.

Finally, the legal and policy "factories" are in constant work. The objects they produce continuously change, reform themselves, aggregate or distinguish each others. The approach to these phenomena by graphs that are either the support for dynamics (Egestedt and Mesbahi, 2010), flows or events, or even whose topology changes is therefore essential.

We can even imagine that these formal approaches based on graph theory will soon be integrated in the design of inter-sectoral integrative policies, thus responding to a growing need for multi-level societies, as evidenced by the everyday news.

7. Acknowledgment

R. Boulet benefits from a post doctoral grant of the CARTAM–SAT project (FEDER 2007-2013 operational program in French Guyana). This study is partly funded (see Sec. 5) by the RTRA *Science and Technology for Aeronautics and Space* (http://www.fondationstae.net/) in Toulouse (France) under the MAELIA project (http://maelia1.wordpress.com/). The yEd Graph editor has been used for producing the figures. Statistical properties of networks have been computed with R and the library igraph (http://www.rproject.org/).

8. References

André, P.; Delisle, C. E. & Revéret, J. P. (2004). *Environmental Assessment for Sustainable Development - Processes, Actors and Practice*, Presses Intern. Polytechnique, École Polytechnique de Montréal, ISBN 2-553-01138-5, Québec, Canada

Bécane, J.-C.; Couderc, M. & Hérin, J.-L. (2010). *La Loi*, Dalloz, ISBN : 978-2-247-08761-7, Paris, France

Berners-Lee, T.; Hendler, J. & Lassila, O. (2001). The semantic web. Scientific American, 284(5), pp. 34-43

Bommarito, M. J. & Katz, D. M. (2010). A Mathematical Approach to the Study of the United States Code (March 25, 2010). Physica A, Vol. 389,19, pp. 4195-4200

Boulet, R. (2011). Introduction d'indices structuraux pour l'analyse de réseaux multiplexes. *Actes 2de Conf. Modèles et Analyse des Réseaux: Approches Mathématiques et Informatique*, October 19-21, Grenoble, France, *in press*

Boulet, R.; Mazzega, P. & Bourcier, D. (2010). Network analysis of the French environmental code, In *AICOL Workshops 2009:* Beijing, China / Rotterdam, The Netherlands, Casanovas, P., Pagallo, U., Sartor, G. & Gianmaria A. (Eds.), LNAI 6237, Springer, Heidelberg, Germany, ISBN 978-3-642-16523-8, pp. 39-53

Boulet, R.; Mazzega, P. & Bourcier, D. (2011a). A network approach to the French system of legal codes - Part I: analysis of a dense Network. Artificial Intelligence & Law, vol.19 (4), 333-355

Boulet, R.; Mazzega, P. & Bourcier, D. (2011b). A network approach to the French system of legal codes - Part II: the role of the weights in a network, *submitted*

Boulet, R.; Mazzega, P. & Bourcier, D. (2012). Réseaux normatifs relatifs à l'environnement : structures et changements d'échelles. In *Politiques Publiques Systèmes Complexes*, Bourcier, D., Boulet R. & Mazzega P. (Eds.), Hermann, Paris, France, *in press*

Bourcier, D. & Mazzega, P. (2007a). Toward measures of legal complexity, *Proc. 11th Intern. Conf. Artificial Intelligence & Law*, Stanford Law School, ACM Press, ISBN 978-1-59593-680-6, New York, USA, pp. 211-215

Bourcier, D. & Mazzega, P. (2007b). Codification, law article and graphs, In *Legal Knowledge and Information Systems*, JURIX, Lodder, A.R. & Mommers L. (Eds.), IOS Press, ISBN:1586038109, pp. 29-38

Bourcier, D.; Boulet, R. & Mazzega, P. (eds.) (2012). *Politiques Publiques Systèmes Complexes*, Hermann, Paris, France, *in press*

Brachman, R. J. & Levesque, H. J. (Eds.) (1985). *Readings in Knowledge Representation*, Morgan Kaufmann, San Mateo, USA (CA)

Brachman, R. J. (1979). On the epistemological satus of semantic networks. Findler, pp. 3-50

Brachmann, R. J. & Levesque, H. J. (2004). *Knowledge Representation and Reasoning*, Morgan Kaufmann Publishers, Elsevier, ISBN: 1-55860-932-6, San Francisco, USA

Brandes, U. & Erlebach, T. (2005). *Network Analysis - Methodological Foundations*, Springer, ISBN: 3-540-24979-6, Berlin, Germany

Casanovas, P.; Noriega, P., Bourcier, D. & Galindo, F. (2007). *Trends in Legal Knowledge - The Semantic Web and the Regulation of Electronic Social Systems*, European Press Acad. Publ., ISBN: 8883980492, Florence, Italy

Chein, M. & Mugnier, M.-L. (2009). *Graph-based Knowledge Representation: Computational Foundations of Conceptual Graphs*, Springer Verlag London, ISBN: 978-1-84800-286-9, London, UK

Clauset, A.; Newman, M. E. J. & Moore, C. (2004). Finding community structure in very large networks, *Physical Review E* 70:066, 111, doi:10.1103/PhysRevE.70.066111

Colizza, V.; Flammini, A., Serrano, M. A. & Vespignani, A. (2006). Detecting rich-club ordering in complex networks, *Nature Physics*, 2, pp. 110-11

Cox, D. J. & Holder, L. B. (2007). *Mining Graph Data*, Wiley & Sons, ISBN: 13 978-0-471-73190-0, New Jersey, USA

CROSSROAD, (2010). A participative roadmap for ICT research in electronic governance and policy modeling – State of the art analysis. D1.2, FP7-ICT-2009-4 SA Project

Delmas-Marty, M. (2007). *Les Forces Imaginantes du Droit III - La Refondation des Pouvoirs*, Le Seuil, ISBN2020912503, Paris, France

Erdös, P. & Rényi, A. (1959). On random graphs, *Publicationes Mathematicae*, 6, pp. 290–297

European Commission, (2004). Commission Report on Impact Assessment: Next steps – In *Support of Competitiveness and Sustainable Development* SEC(2004)1377 of 21 October 2004

Farquhar, A.; Fikes, R., Pratt, W. & Rice, J. (1995). *Collaborative Ontology Construction for Information Integration*, Knowledge Systems, AI Laboratory Department of Computer Science, KSL-95-63

Freeman, L. C. (1979). Centrality in social networks: conceptual clarification, *Social Networks*, 1(3), pp. 215–239

Gangemi, A.; Guarino, N., Masolo, C. & Oltramari, A. (2001). Understanding top-level ontological distinctions, *Proc. of IJCAI 2001 Workshop on Ontologies and Information Sharing*, Gómez Pérez, A., Gruninger, M., Stuckenschmidt, H. & Uschold, U. (Eds.), Seatle, USA

Genesereth, M. R. & Fikes, R. E. (1992). *Knowledge Interchange Format*, Version 3.0 Reference Manual. Technical Report Logic-92-1, Computer Science Department, Stanford University, USA

Grüber, T. R. (1992). *Ontolingua: A Mechanism to Support Portable Ontologies*. Technical Report KSL 91-66, Knowledge Systems Laboratory, Stanford University, USA

Grüber, T. R. (1995). Toward principles for the design of ontologies used for knowledge sharing, *Intern. J. Human-Computer Studies*, 43 (5-6), pp. 907–928

Grüber, T. R. (2009). Ontology, In the *Encyclopedia of Database Systems*, Liu, Ling; Özsu, M. Tamer (Eds.), Springer-Verlag, *ISBN* 978-0-387-35544-3, Berlin, Germany

Guarino, N. (1998). Formal Ontology in Information Systems. In *Formal Ontology in Information Systems*, Guarino N. (Ed.) *Proceedings of FOIS'98*, Trento, Italy, June 6-8, 1998. IOS Press, ISBN: 9051993994, Amsterdam, The Netherlands, pp. 3-15

Hassenteufel, P. (2008). *Sociologie Politique : l'Action Publique*, Armand Colin, coll. U Sociologie , *ISBN*-10: 2200019858, Paris, France

Henry, N. (2004). *Public Administration and Public Affairs* (9th Ed.), Prentice-Hall Inc., ISBN10: 0-13-222297-3, Upper Saddle River, NJ USA

Howlett, M. & Ramesh, M. (2003). *Studying Public Policy – Policy Cycles and Policy Sub-Systems*, Oxford Univ. Press, *ISBN*-10: 0195417941, Oxford, UK

Jungnickel, D. (1999). *Graphs, Networks and Algorithms*, Algorithms and Computation in Math. Vol.5, Springer, ISBN 3-540-63760-5, Berlin, Germany

Kelsen, H. (1960). *Pure Theory of Law* (2d. Ed.), M. Knight, trans. (1967), University of California Press, Berkeley, USA

LEMA, (2006). *Loi n°2006-1772 du 30 décembre 2006 sur l'eau et les milieux aquatiques*, JORF n°303 (31/12/2006), texte n°3, p.20285 sq. http://www.legifrance.gouv.fr/

Louka, E. (2006). *International Environmental Law*. Cambridge Univ. Press, ISBN-13: 978-0-521-86812-9, Cambridge, UK

Mazzega, P.; Bourcier, D. & Boulet, R. (2009). The network of French legal codes. *Proc. 12th Intern. Conf. Artificial Intelligence and Law*, ACM 2009, ISBN 978-1-60558-597-0, Barcelona – Spain, June 8-12, pp. 236-237

Mazzega, P.; Bourcier, D., Bourgine, P., Nadah, N. & Boulet, R. (2011). A complex-system approach: legal knowledge, ontology, information and networks. In *Approaches to*

Legal Ontologies: Theories, Domains, Methodologies, Sartor, G., Casanovas, P., Biasiotti, M. A. & Fernández-Barreira, M. (Eds.), Springer, ISBN 978-94-007-0119-9, Berlin, Germany, pp. 117-132

Mesbahi, M. & Egestedt, M. (2010). *Graph Theoretic Methods in Multi-Agent Networks*, Princeton Series in Applied Math., Princeton Univ. Press, ISBN: 9780691140612, Princeton, USA

Minsky, M. (1975). A Framework for Representing Knowledge, In *The Psychology of Computer Vision*, Winston, P. H. (Ed.), McGraw-Hill, USA

Ost, F. & van de Kerchove, M. (2002). *De la Pyramide au Réseau? Pour une Théorie Dialectique du Droit*, Publ. Facultés Univ. Saint-Louis, ISBN2802801538, Bruxelles, Belgium

Ostrom, E. (2005). *Understanding Institutional Diversity*, Princeton Univ. Press, ISBN: 9780691122380, Princeton, USA

Pagé, C. & Terray, L. (2011). New fine-scale climate projections on France for the 21st century: scenarios SCRATCH2010. Climate Modelling and Global Change Technical Report TR/CMGC/10/58 (in french). Centre Européen de Recherche et de Formation Avancée en Calcul Scientifique (CERFACS), Toulouse, France

PDM AG, (2010). *Programme de Mesures du Bassin Adour-Garonne 2010-2015*, Comité de Bassin AG / Ministère de l'écologie, de l'énergie, du développement durable et de la mer, adopté le 16 nov. 2009, http://www.eau-adour-garonne.fr/

Pons, P. & Latapy, M. (2005). Computing communities in large networks using random walks. *Journal of Graph Algorithms and Applications*, 10(2), pp. 191–218

Quillian, M. R. (1968). Semantic memory, In *Semantic Information Processing*, Minsky M. (Ed.), M.I.T. press, Cambridge, USA (MA), pp. 216-270

Sartor, G.; Casanovas, P., Biasiotti, M. A. & Fernández-Barreira, M. (Eds.) (2011). *Approaches to Legal Ontologies: Theories, Domains, Methodologies*, Springer, ISBN 978-94-007-0119-9, Berlin, Germany

Schneider, Ch. J. & Urpelainen, J. (2012). Distributional conflict between powerful states and international treaty ratification. International Studies Quarterly, *forthcoming*.

SDAGE AG, (2010). Schéma Directeur d'Aménagement et de Gestion des Eaux du Bassin Adour-Garonne 2010-2015, Comité de Bassin AG / Ministère de l'écologie, de l'énergie, du développement durable et de la mer, adopté le 16 nov. 2009, http://www.eau-adour-garonne.fr/

Sowa, J. F. (1984). *Conceptual Structures, Information Processing in Mind and Machine*, Addisson-Wesley, ISBN: 0201144727, USA

von Luxburg, U. (2007). A tutorial on spectral clustering, *Statistics and Computing* 17(4):395416, URL http://arxiv.org/abs/0711.0189, 0711.0189.

Watts, D. J. (2003). *Small Worlds: The Dynamics of Networks between Order and Randomness*, Princeton University Press, ISBN: 9780691117041, Princeton, USA

WFD, (2000). *Directive 2000/60/EC of the European Parliament and of the Council of 23 October 2000 establishing a framework for Community action in the field of water policy*, Available from: http://eur-lex.europa.eu/en/

Zhou, S. & Mondragon, R. J. (2004). The rich-club phenomenon in the internet topology, *IEEE Communications Letters*, 8(3), pp. 180–182

12

A Dynamic Risk Management in Chemical Substances Warehouses by an Interaction Network Approach

Omar Gaci[1] and Hervé Mathieu[2]
[1]ISEL, Le Havre University
[2]LITIS - ISEL, Le Havre University
France

1. Introduction

Supply chain is a set of activities involving a group of commercial actors to create a product or a service to satisfy a customer demand. The actors are the ones who form the supply chain, they are suppliers, transporters, manufacturers, distributors, retailers, customers. The objective of every supply chain is to maximize total supply chain profitability.

The Supply Chain puts in interaction a set of entities to provide to the final client the right product (or service) at the right time. Raw material suppliers, manufacturers of parts and components, assemblers, original equipment manufacturers, distributors, retailers, and customers are the main interacting entities of supply chain (SC) systems (Forrester, 1961).

In this chapter, we model Supply Chain as Complex Adaptive System (CAS) (Holland, 1996). CAS postulates that the activities of the constituting entities contribute to a specific emergence which corresponds to a global behaviour. Thus, the system is composed by active and adaptive intelligent agents. Their behaviours, interactions and adaptations lead to the emergence of the system behaviour.

We propose to study activities of storages in a warehouse of chemical substances. Then, this warehouse is subject to restriction in business processes executed every day: operators must respect a segregation strategy which consists in avoiding any mixing of incompatible chemicals. To reproduce the actions of forklift operators, we propose a Multi-Agent System which is the support of CAS modelling. Then, during agent movements for handling pallets from their reception into their storage locations, we define a dynamic graph where the vertices represent agents in activities and edges measure the distance between agents. The study of this dynamic graph shows that the average mean distance remains weak meaning that agent are often close each other. From this observation, we deduce a strategy for a dynamic risk management that gives the priority to agents whose betweenness is superior to the other agents that handle pallets of incompatible chemicals.

This chapter is organised as follows. In section 2, we present the main features of a Complex Adaptive Supply Chain and notably the existing support to simulate such a Supply Chain through Multi-Agent System. In section 3, we describe the existing solutions to reproduce

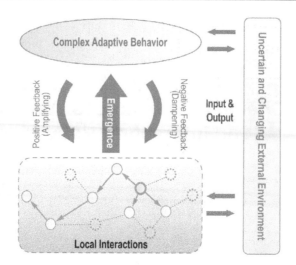

Fig. 1. Emergence from local interactions in Complex Adaptive Systems.

the activities of storages in a warehouse of chemical substances. In particular, we highlight the JADE framework for Multi-Agent simulations. In section 4, dangerous goods in logistics are studied and also the current regulation that warehouse must comply with. In section 5, we present our solution to reproduce the activities of storages in the studied warehouse by implementing a Multi-Agent system. In section 6, we study the dynamic graph resulting from agents handling actions and we deduce a dynamic risk management strategy.

This work is funded by German BMBF and French ANR as part of ReSCUeIT project.

2. Complex Adaptive Supply Chain

The theory of Complex Adaptive Systems (CASs) is presented by Holland (Holland, 1996) as a new paradigm to study the organizations and the dynamics of multi-scale systems whose evolution and adaptability leads to a global behaviour. A CAS can be considered as a multi-agent system with seven basic elements. According to Holland, the first four concepts are aggregation, nonlinearity, flow and diversity. They represent the characters of agents and influence the adaptability and the system evolution. The last three concepts, tagging, internal models and building blocks, are specific mechanics for agents to communicate with each other and also with the environment. The environment is itself subject to evolution notably because of the agent interactions which compete or cooperate from a same resource or for achieving a specific goal. As well, since the environment changes, the agents' behaviour evolve as consequence.

The main remarkable property which characterizes a CAS is the emergence of highly structured collective behaviour over time from the interactions of simple entities (Holland, 1996). The emergence of a complex adaptive behaviour from the local interactions of the agents is illustrated in Fig. 1. Then, the CAS and its environment evolve in the same time in order to maintain themselves in a state of quasi-equilibrium.

Considering a CAS consists in studying non-linear phenomena, non-exhaustive knowledge, a large state numbers and dynamic changes in environment. The main challenges when we reproduce a CAS are to produce a global behaviour by emergence under unpredictable conditions.

2.1 Complex Adaptive Systems as Multi-Agent Systems

In a Complex Adaptive System, CAS, a global behaviour emerges over time into a coherent form, adapting and organizing themselves without any singular entity controlling or managing the global structure or node interactions (Holland, 1996).

Complex Adaptive Systems are commonly implemented and simulated by Multi-Agent System, MAS, (Julka et al., 2002; Kwon et al., 2006; Swaminathan et al., 1997) which represents a general and flexible framework to describe and model autonomous systems including their interactions. An agent is basically a self-directed entity with its own goals and has a means to interact or to communicate with other agent.

2.1.1 Multi-Agent Systems

A MAS is formed by a network of computational agents that interact and typically communicate with each other.

The approach described by MAS consists in representing explicitly the individuals or the entities which compose the studied population. The system can then be ecological, social economic, etc. The goal is to produce a model for the entities, for the environment and for the mutual interactions. When the entities and their interactions are modelled, it remains to study the evolution of the relation through the simulation of their collective behaviour.

The understanding of the global MAS dynamic is viewed according to two levels, the microscopic (the study of the individual dynamics) and the macroscopic (the observation of the collective behaviour resulting of the entities interactions).

In (Conte, 1999), the authors propose two conceptual approaches deduced from the modelling of social phenomena:

* The top-down approach, which enables to deduce the microscopic phenomena, the goals or the individuals' motivations starting from the macroscopic observations;
* The bottom-up approach in which the hypothesis are established on the individual behaviours, their motivation or their way of interacting. The observation of their collective behavior is then compared to the macroscopic phenomena observed in the modelled system to eventually discuss the hypothesis formulated at a microscopic level (Epstein & Axtell, 1996). The bottom-up approach is specific to the most individual based models which propose such models to explain or characterize observed collective behaviours.

2.1.2 The agent properties

The agents considered in MAS are used in a broad variety of applications and are defined by the following way (Ferber, 1999):

The term 'agent' denotes a hardware or (more usually) software-based computer system, that has the following characteristics (Casterfranchi, 1995):

- *Autonomy*: an agent acts without any intervention from its environment and possesses rules to control its action and internal states;
- *Social ability*: an agent interacts and communicates with other agents using a specific agent-communication language;
- *Reactivity*: an agent perceives its environment and is able to answer to any solicitations;
- *Pro-activeness*: an agent is not only reactive to a stimulus from its environment, it is also able to exhibit goal-directed behaviour by taking the initiative.

2.1.3 The type of agents

Agents are defined by their capacities and according to these properties, different levels of complexity characterize agents. Such complexity depends on the task that agents have to carry out and on the environment surrounding them. In (Ferber, 1999), agents are classified according to their architectures:

- *Simple reflex agents*: These agents are basic because their actions depend on stimulus. Their act are then subject to specific conditions. The past is not considered and none memory influence the present reactions.
- *Model-based reflex agents*: These agents cannot perceive their whole environment but keep track of their environment they cannot currently observe. Then, they possesses an internal representation of their environment called 'model of the world' to evaluate the environment evolution and the impact of the agent's actions on this environment. These agents select their action according to condition-action rules. The conditions only depends on the model of the world, and not on the current perception of the environment.
- *Model-based, goal-based agents*: These agents have goals describing desirable situations to choose an action because the current state of the model of the world is not always enough to select an action efficiently. The model of the world represents a state from which the agents evaluate how the world would be after an action. The action is chosen in order to the agent goals are satisfied and the model of the world state is a parameter taken into account.

2.1.4 Applications of Multi-Agent Systems

MAS are commonly exploited to model and simulate one of the three followings types of applications:

- *MAS for studying complexity*. These studies regroup social models such as the segregation model of Schelling (Schelling, 1971) artificial life simulation with the Sugarscape (Epstein & Axtell, 1996) and Reynold's Boids models (Reynolds, 1971) and also logistics models (for example traffic simulations (Burmeister et al., 1997)). These models are built with simple reactive agents and a set of rules without any need of resource planning or coordination. The simulations are monitored relying on qualitative measures (emergent communities, emergent flocking, emergent behaviour) and/or quantitative (average generation, average agent movements, average awaiting time). These models are then studied as well from a top-down approach than with a bottom-up approach.
- *MAS for studying Distributed Intelligence*. These studies are relative to planning (Pollack & Ringuette, 1990) and particularly cognitive social interactions Doran (n.d.); Gilbert (2005); Sun (2001). The main goal is to reproduce human cognition through

cognitive agents (Sloman & Logan, 1999). These developed models use complex, situated and communicating agents to study the behaviour of cognitive formalism (Taatgen et al., n.d.; Wray & Jones, n.d.).

- *Application development with MAS.* Existing toolkits provide technical tools to develop software agents described in (Jennings et al., 1998). Software agents are then Semantic Web agents, Beliefs-Desires-Intention (BDI) agents in expert systems, or agents for network metamanagement. These toolkits include a development environment to implement MAS, it can be considered equivalent to a simulation engine.

Further, in this chapter, software agents are used and the JADE platform is used as framework for the development.

3. Multi-Agent System to study warehouse activities

The activities of modelling and simulating offer applications in scientific and industrial fields. These works improve the understanding and the reliability of design of various systems.

In the context of supply chain, the study of warehouse activities is motivated by the following goals:

- test by a software a virtual version of a warehouse before implementing and using the real system;
- collect information to support discussion with the customer;
- simulate the warehouse activities to improve business or security procedures;
- generate reproducible error situations.

In this chapter, we are interested in simulating storage activities in a warehouse of dangerous goods to evaluate the segregation policies efficiency and to propose a reliable risk management to maintain these segregation strategies.

3.1 Existing softwares in warehouse simulations

Over the years, different tools have been developed to help designers and users to model and simulate warehouse activities. Existing tools can be divided in three groups: GUI-based simulation softwares, framework libraries and specialized programming languages (Colla & Nastasi, 2010).

There is few simulation tools designed for the application of supply chain activities. Among them, we can cite the commercial tools eM-Plant. It can be used for visualization, planning and optimization of production and logistics. FlexSim (*Flexsim Simulation Software*, n.d.) is another commercial software which enables fast and easy modeling, clear visualization as well as reuseability of models.

The agent-based approach appears to be a powerful tool for the development of complex systems and is exploited in industrial applications (Weiming et al., 2000). This approach is used in many fields such as manufacturing, process control, telecommunication, air traffic control, transportation systems, information management, electronic commerce, etc.

Among the existing agent-based applications for the simulation of supply chain activities, we can cite Repast, SeSAm, NetLogo, SDML or AnyLogic.

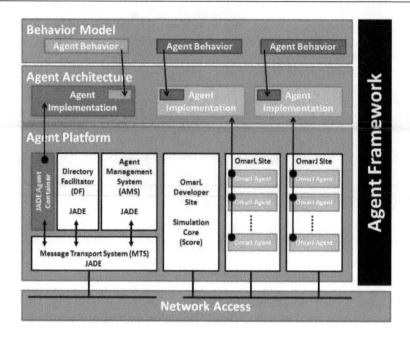

Fig. 2. Description of Agent Framework.

3.2 Existing frameworks for Multi-Agent warehouse simulations

It exists many different types of frameworks dedicated to the agent development. They are built from different theories and principles and allows then their classification. The definition of an agent framework is the following: an agent framework is a dedicated structure or platform to the development of software agents and based on a specific technical architecture.

An agent framework covers a set of missions relative to agents development, platform development, agent architecture and agent behaviour models as shown by Fig. 2. Then, an agent development platform is a structure which encompasses and support the entire life cycle of agents and provide in the same time a communication interface for agent interaction. This agent development platform commonly provides an API that defines the manner an agent communicate within the platform (Bellifemine, Caire, Trucco & Rimassa, 2007). As well, the agent architecture constitutes itself a framework for creating behaviour models. A behaviour model represents the architecture content and usually represents different forms of knowledge. The behaviour can be viewed as the result of the architecture and its content (Lehman et al., 2006).

The Java Agent Development Environment (JADE) is an agent platform used further in this chapter to implement a Multi-Agent System. JADE is a platform for the creation of MAS and contains a message transport system (MTS). This MTS, constitutes a network interface for developing distributed agent networks. As well, an Agent Management System (AMS) is available and allows the supervising of agent access control to the MTS and a directory facilitator (DF) for creating distributed services. In JADE, an agent is an instant that runs in the agent platform, then the agent has a determined life cycle by the AMS. Each agent is able to communicate with the other and possesses a queue for sending and receiving messages. An agent instance represents a container for the agent internal structure. The JADE agent platform is a middle-ware that complies with the specifications of the Foundation for Intelligent Physical Agents (FIPA) (Bellifemine, Caire, Trucco & Rimassa, 2007).

3.3 JADE platform for warehouse activities simulations

JADE is a software development framework fully implemented in JAVA language aiming at the development of multi-agent systems and applications that comply with FIPA standards for intelligent agents (Bellifemine, Caire & Greenwood, 2007). JADE is an agent framework and provides then a set of technical features to the development of MAS such as:

- *A distributed agent platform.* The platform can be easily shared and hosted in different machines when each machine possesses its own Java Virtual Machine;
- *FIPA-Compliant agent platform.* This means that the platform provides a set of functionalities such as Agent Management System, a Directory Facilitator and an Agent Communication Channel;
- *Communication with ACL messages.* The standard ACL ensures efficiently in the message transport between agents.

Communication of agents consists in sending and receiving messages, the FIPA ACL language is used to represent the messages. Each agent possesses an incoming message box and messages can be blocking or nonblocking during a determined blocking time. As well, JADE offers the possibility of filtering messages: it is possible to utilize advanced filters relative to different fields of the incoming message such as sender or ontology.

To build agent conversations, FIPA defines a set of standard interaction protocols such as FIPA-request and FIPA-query that can be exploited as standard for agent communication. When a conversation starts between two agents, JADE distinguishes two roles: the initiator who is the agent that starts the conversation and the responder who communicates with the previous one. This protocol architecture implies that the initiator sends a message and the responder can potentially reply by refusing the message indicating the incapability to continue with the conversation. The responder can also answer with a agreed message indicating that the communication between the two agents is established and can continue. After receiving a message, the responder performs potentially an action and must send back a message to describe such an action. In case that the action has failed, a failure message indicates that the action was not successful. JADE provides behaviour for initiator and responder roles according to FIPA interaction protocols. Then, the classes *AchieveREInitiator* and *Responder* provides homogeneous implementation of interaction protocols with methods for handling the different communication phases.

In JADE, agents actions or missions are implemented by the implementation of *behaviours*. These behaviours are defined as threads that can be composed or not, and allows agents to achieve their intentions. Such behaviours can be initialized, suspended and spawned at any given time. Then, the agents possess an action list that is executed through their behaviours. The JADE platform uses one thread per agent and not one thread per behaviour due to resource concerns (the number of running threads is limited). As well, a scheduler (unreachable for developers) organizes via a round-robin strategy the behaviours already created and instantiated in the queue. The coding of a behaviour offers the possibility of releasing the execution control when blocking mechanisms are used. The behaviours are executed in the method action().

The behaviour of agents is defined by a *Behaviour* class that can be specialized to defines a set of other behaviours. A behaviour is composed by several methods so that it is possible to describe the different state transitions. From this root behaviour, children behaviours can be deduced and notably the *SimpleBehaviour* and *CompositeBehaviour*. Behaviours that specialize or descend from *SimpleBehaviour* represent atomic simple tasks that can be executed several times according to the developer coding. As well, behaviours from *CompositeBehaviour*, are able to use multiple behaviours according to the children behaviours. Then, the agent tasks are executed not directly through the current behaviour but inside its children behaviours. For that purpose, the *FSMBehaviour* class executes the children behaviours. The *FSMBehaviour* class is able to maintain the transitions between states and to select the state coming after the current one. It is possible to register some of the children of an *FSMBehaviour* as final states. This type of behaviours terminates once one of its children has finished its execution.

4. Dangerous goods in logistics

A good is considered as dangerous when it may present a danger on the population, the environment or on the infrastructures according to its physicochemical properties or because of the reactions it can imply. A dangerous good can be flammable, toxic, explosive, corrosive or radioactive. According to the new CLP regulation, dangerous goods are considered as chemical substances in the European Union.

4.1 Dangerous goods identification

Considering the important number of substances, there is a clear need for dangerous goods classification. Amongst the existing classification of dangerous goods, the following distinctions exist:

- chemical family (acid, alcohol, amide, etc.);
- chemical reaction (oxidation, reduction, combustion).

We remark that the vocabulary becomes quickly specialized. To avoid this technical aspect, the dangerous goods are described in function of their reactions. Thus, the danger that represents the manipulation of dangerous goods depends on the properties of each product. Some goods represent only one risk whereas others regroup several.

The CLP (Classification, Labelling, Packaging) regulation is relative to the chemical substances imported or commercialized in the European Union. This regulation entered into force in January 2009 and will be totally applied in 2015.

4.1.1 Obligations under CLP

CLP provides a global obligation for all suppliers in the supply chain to cooperate. This cooperation is necessary to make the different suppliers meet the requirements for classification, labelling and packaging.

4.1.2 Terminology

A new terminology is used, terms of existing regulation are kept whereas news are adopted. The term substance is used to designed hazardous material and the transformation of these substances into a new one is called mixture.

As well, the properties of substances are described according to three properties: physicocochemical, toxicological and ecotoxicological. According to these three criterion, the definition of hazard classes helps to classify a substance. Then, a hazard class defines the nature of a hazard, it can be physical, on health or on the environment.

4.1.3 Classification of substances

CLP possesses specific criteria of classifications that are rules that allow associating a substance to a class of hazard or a category in this class. In particular, the classification process is based on the substance concentrations to establish the effects of those substances on the health and the environment.

CLP defines three hazard classes and 28 categories, such as:

- 16 categories for physical hazards;
- 10 categories for health hazards;
- 2 categories for environmental hazards.

For example, the physical hazards regroup explosives, flammable gases, solids, aerosols, liquids. The health hazards are relative to acute toxicity, skin corrosion, irritation and sensitization. The environmental hazards address hazardous to the aquatic environment and hazardous to the ozone layer.

4.1.4 Labelling

A substance contained in packaging should be labelled according to the CLP rules with the following information (called labelling elements):

- the name, address and telephone number of the supplier of the substance;
- the quantity of the substance in the packages;
- hazard pictograms;
- signal word;
- hazard statements;
- appropriate precautionary statements;
- supplemental information.

A substance contained in packaging is labelled according to the CLP rules and contains a set of information such as name of the supplier of the substance, quantity of the substance in the packages or hazard pictograms, see Fig. 3.

The CLP regulation helps then the identification of chemical substances through the supply chain since it provides a standard framework for the classification, the labelling and the packaging of substances.

Fig. 3. Pictograms used in CLP regulation.

4.2 Dangerous goods storage

Among dangerous goods, products can react violently when they are in contact. For these reasons, they must be stored in separate places. The strategy of storage consists in avoiding incompatible products to be neighbours. To avoid any storage of incompatible goods and risks of chemical reactions in case of wrong manipulation, segregation policies are established. Fig. 4, summarizes the incompatibilities between chemical substances.

Segregation policies in dangerous good warehouses consist in storing products according to their physicochemical properties. This strategy is static and doesn't take into account the possible movements of incompatible goods (by forklifts for example) that can be present in the same place at the same time.

The segregation can be achieved by the use of an impervious barrier or by a separation distance sufficient to prevent mixing. The segregation policies are also subject to constraint storages. According to the nature of goods, specific storage conditions must be respected. Among storage constraints, we can cite the most obvious such as storage conditions (humidity, heat and light). The respect of these constraints is ensured by safety equipments: sprinkler, smoke detector, particles detector or temperature probe.

Consequently, to make a warehouse of dangerous goods secured, different types of safety equipments are needed and a reliable segregation is used. In this chapter, we propose to simulate such a warehouse and to study the emergent collective behaviour of the MAS constituted by the warehouse actors.

Danger Code	F	F+	T	Xi	O	Xn	N	C
F	+	+	-	+	-	+	-	-
F+	+	+	-	+	-	+	-	-
T	-	-	+	+	-	+	-	-
Xi	+	+	+	+	-	+	+	-
O	-	-	-	-	+	-	-	-
Xn	+	+	+	+	-	+	-	-
N	-	-	+	-	-	-	+	-
C	-	-	-	-	-	-	-	+

Fig. 4. Identification of compatibilities between dangerous goods. The letter F means inflammable, F+ means very inflammable, T means toxic, Xi means very irritant, means O oxidizing, Xn means noxious, N means polluting and C means corrosive.

5. Simulation of warehouse activities by a Multi-Agent System implemented with JADE

Agent considered represent forklift drivers and the warehouse structure is then the agent environment. Fig. 5 shows the warehouse architecture which is composed of a forklift base, corridors, docks where pallets are temporally stored and five racks.

5.1 Forklift agent

Forklift agents are simple reactive agents that are positioned in their base and wait for a message from the central warehouse scheduler. This scheduler is actually a random generator which creates truck arrivals and sends messages to forklift driver agents so that they go to docks to unload the truck. Once the truck is completely unloaded, forklift driver agents continue their actions and store pallets in their rack position.

As shown by Fig. 6, agents react and communicate through messages. Firstly, agents are in their base and when they receive a message *GoToDock* they receive also the dock number and they consecutively move according to the *moveToDock(dockNum)* method. The agent motion follows warehouse corridors and this method provides to agents the set of corridors to use in order to reach the dock number *dockNum*. When the agents is in position, he confirms his position to the centralized scheduler and replies a message *atDock-DockNum*. This means that he is operational and the scheduler communicates with him to ask him to begin the unloading with an *Unload* message. If the scheduler has sent another message type, the agent would move back to the base. To unload the truck, the agent executes the *unloadTruck* method and confirms the end of handling operations with a message *EmptyT-dockNum*. Once more, the scheduler can choose to call back the agent to the base or to send him a message *Storage* so that the agent uses the method *storePalletsFrom(dockNum)*. This last method indicates to the

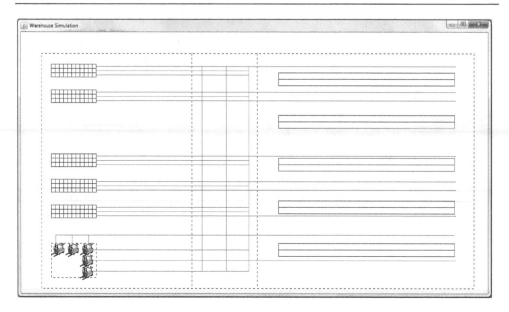

Fig. 5. Representation of the studied warehouse. Agents are located in their base and follow corridors to go to docks or racks.

agents the corridors to follow until the storage place in rack. When, the agent finishes the storage of pallets, he sends a message *StoredP-DockNum* and move back to his base.

The forklift driver agents evolve in a warehouse which represents their environment. They interact with a set of objects enumerated in the class diagram presented in Fig. 7. As shown, a forklift agent is a software agent defined according to an *ID* which is typically a number. His position is monitored in two dimensions and the current *Corridor* where he is evolving is given. As well, the *status* variable provides a means to know if the agent is in activity or if he is waiting in the base. In the class *Pallets*, the *hazardType* attribute gives the type of dangerous goods present on the pallet and it is the same for *Racks* that stores only restricted types of dangerous goods.

6. An interaction network approach for a dynamic risk management

The interaction network approach proposed in this section consists in monitoring in real-time the forklift agent movements and to detect a risk of incompatible chemicals mixing. To achieve such a goal, the warehouse is viewed as a dynamic graph where forklift agents who are active in the warehouse represent vertices that can be removed when they move back to their base. Edges link the vertices which represent agents and are weighted by the euclidean distance between agents in the warehouse. By this way, it is possible to consider a dynamic graph that puts in interaction forklift agents whose edges represent distance between them. The goal is then to detect the risk of incompatible chemicals mixing when the distance between agents is insufficient.

```
addBehaviour(new CyclicBehaviour(this) {
private static final long serialVersionUID = 1L;

public void action() {
int dockNum;
String msgType1, msgType2, msgType3;
ACLMessage msg1, msg2, msg3;
ACLMessage rep1, rep2, rep3;

// Waiting message from scheduler
msg1 = receive(MessageTemplate.MatchPerformative(ACLMessage.INFORM));
if (msg1 != null){
  msgType1 = msg1.getContent().substring(0,8);
  if (msgType1.equals("GoToDock")){

    dockNum = Integer.parseInt(""+msg1.getContent().substring(8,9));

    moveToDock(dockNum);    //move to dock
    rep1 = msg1.createReply();
    rep1.setContent("atDock"+dockNum);
    send(rep1);             // moved at dock

    msg2 = receive(MessageTemplate.MatchPerformative(ACLMessage.INFORM));
    if (msg2 != null){
      msgType2 = msg2.getContent().substring(0,8);
      if (msgType2.equals("Unload__")){

        unloadTruck(dockNum);   //unload pallets from a truck
        rep2 = msg2.createReply();
        rep2.setContent("EmptyT"+dockNum);
        send(rep2);             // pallets unloaded from the truck
        }

      msg3 = receive(MessageTemplate.MatchPerformative(ACLMessage.INFORM));
      if (msg3 != null){
        msgType3 = msg3.getContent().substring(0,8);
        if (msgType3.equals("Storage_")){

          storePalletsFrom(dockNum);// store pallets from a dock
          rep3 = msg3.createReply();
          rep3.setContent("StoredP_"+dockNum);
          send(rep3);            // pallets stored from the dock
        }
       }
      }
     }
goToBase();
}
```

Fig. 6. Behaviour of forklift agents. They react after receiving messages and confirm their action by replying.

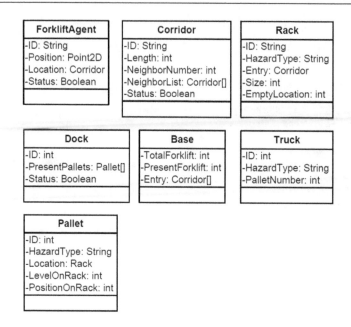

Fig. 7. Classes used to reproduce the activities of storage in a warehouse of chemical substances.

6.1 Dynamic graphs

Many systems, both natural and artificial, can be represented by networks, that is by sites or vertices bound by links. The study of these networks is interdisciplinary because they appear in scientific fields like physics, biology, computer science or information technology. The purpose of these studies is to explain how elements interact inside the network and what are the general laws which govern the observed network properties.

From physics and computer science to biology and the social sciences, researchers have found that a broad variety of systems can be represented as networks, and that there is much to be learned by studying these networks (Broder et al., 2000). Indeed, the study of the Web (Albert et al., 1999), of social networks (Wasserman & Faust, 1994) or of metabolic networks (Jeong et al., 2000) are contribute to put in light common non-trivial properties to these networks which have *a priori* nothing in common. The ambition is to understand how the large networks are structured, how they evolve and what are the phenomenon acting on their constitution and formation (Watts & Strogatz, 1998).

Nevertheless, to study the dynamic of a phenomenon through a graph, we need tools able to describe the graph topology evolution over time. The works relative to random graphs (Erdos & Rényi, 1959) provide a generic dynamic model which describe graphs whose edges are added according to a specific probability.

More recently, the interest for dynamic graphs has increased notably because of their potential application in communication, urban traffic or social sciences. The dynamic graphs allow studying the graph topology evolutions relying on dynamical metrics able to describe the graph properties when it evolves over time.

Now, we give some graph theory definitions to propose a definition of dynamic graphs.

A graph G is formally defined by $G = (V; E)$ where V is the finite set of vertices and E is the finite set of edges each being an unordered pair of distinct vertices.

Let f be a function defined on the vertex set as $f : V \rightarrow N$, then the triple $G = (V; E; f)$ is a node weighted graph. As well, let g be the function defined on the edge set as $f : E \rightarrow N$, the triple $G = (V; E; g)$ is an edge weighted graph.

In (Harary & Gupta, 1997), the authors classify the dynamic graphs as a function of the graph evolution:

- Node dynamic graphs, the vertex set V changes over time
- Edge dynamic graphs, the edge set E is modified over time
- Node weighted dynamic graphs, the f function evolves over time
- Edge weighted dynamic graphs, the g function varies over time

6.2 Graph metrics

Different graph measures allow characterizing graphs. Here, the proposed metrics provide measures for global description and also for individual vertices so that it is possible to identify the influence of a vertex in the modelled warehouse.

6.2.1 Distance and diameter

The distance in a graph $G = (V, E)$ between two vertices $u, v \in V$, denoted by $d(u, v)$, is the length of the shortest path connecting u and v.

A graph diameter, D, is the longest shortest path between any two vertices of a graph:

$$D = \max\{d(u, v) : u, v \in V\}$$

The mean distance is defined as the average distance between each couple of vertices:

$$L = \frac{2}{n(n-1)} \sum_{u,v \in V} d(u, v)$$

6.2.2 Mean degree

A degree of a vertex u, k_u, is the number of edges incident to u. The mean degree, z, of a graph G is defined as follows:

$$z = \frac{1}{n} \sum_{u \in V} k_u = \frac{2m}{n}$$

6.2.3 Node betweenness

The betweenness of a node is defined as the total number of shortest paths between pairs of nodes that pass through this node. It measures the influence of a node in a network. The betweenness of a node t, denoted $B(t)$ is defined as follows:

$$B(t) = \sum_{u \neq v, u \neq t, v \neq t} \frac{\sigma_{uv}(t)}{\sigma_{uv}}$$

where σ_{uv} is the number of shortest paths between the nodes u and v, and $\sigma_{uv}(t)$ is the number of shortest paths between u to v that pass through t.

6.3 Simulation and results

We assume that the warehouse is well dimensioned and at last one agent is available to perform a truck unloading or a pallet storage. Once a single or several agents react, they perform handling operations according to their behaviours. Then, agents are located in their forklift base and wait for a truck arrival. When a truck is in position, agents react and move into a dock. Another reaction is needed in order to agents unload the truck. The movements of agents follow the warehouse corridors. The time spent by agents to unload a truck or to store pallets is the average time observed in the real warehouse. The objective is to simulate a behaviour close to the reality.

In this case study, we consider that the warehouse stores three types of chemical substances denoted A, B and C. Each product must be stored only in its rack location and the segregation consists in avoiding that a product is stored in another rack. We consider 6 forklift agents and 10 trucks with a cargo of 33 pallets by truck. We launch a simulation and we study the dynamic graph resulting for agents activities.

Fig. 8 shows the evolution of the mean distance denoted l and the diameter, D. It appears that the means distance evolves between 30 and 150 which is the consequence of the warehouse dimensions. As well, the diameter being an upper bound of distances in interaction networks, we expect that the mean distance l will be lower than D. The mean degree is studied in Fig. 9 and shows that it evolves between 1 and 6. This means that at least two agents are in activities in the warehouse and when they are all outside their base, the mean degree will be 6.

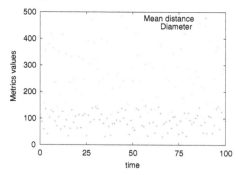

Fig. 8. Mean distance and diameter of the resulting graph from agent activities.

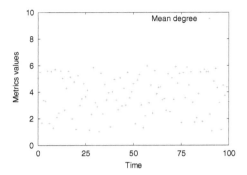

Fig. 9. Mean degree of the resulting graph from agent activities.

This first study about the dynamic graph resulting from agents handling operations put in evidence that they are not all present in same time in the warehouse. The average distance between agents is still weak in front of the upper bond expressed by the diameter.

Our goal is to develop a dynamic risk management strategy to maintain a segregation during the agent movements. In front of results presented above, we deduce a strategy presented in Algorithm 1. Then, when two agents are present in the same corridor, the type of handled goods determines if these forklift agents can share this corridor. In case that incompatible products are transported by agents, a topological measure is exploited, typically the node betweenness, to determine the priority between these agents. We consider that the agent

---

Algorithm 1: Algorithm to maintain a dynamic segregation between agents

---

Input:
F_a: set of forklift agents
Data:
f_a, f_a': forklift agent
$B(f_a)$: betweenness of the current agent
$HazType(f_a)$: hazard type of chemicals handled by the current agent
$Corr(f_a)$: corridor length of the current agent
$reorienteAgent(f_a)$: current agent is reoriented into another corridor

begin
  foreach $f_a \in F_a$ **do**
    foreach $f_a' \in F_a$ **do**
      if $dist(f_a, f_a') < Corr(f_a)$ **then**
        if $HazType(f_a) and HazType(f_a')$ are incompatible **then**
          if $B(f_a) > B(f_a')$ **then**
            $reorienteAgent(f_a')$
          else
            $reorienteAgent(f_a)$

---

whose betweenness is superior, has the priority and the other agent is reoriented into another corridor. If the next corridor is in the same configuration, the agent will change again until be in presence of incompatible chemicals. Therefore, the dynamic risk management strategy is defined as a prevention of incompatible chemicals crossing in corridors. Any risks of crossing or mixing is mitigated by the routing of agents into another corridor when this agent possesses a weaker betweenness that the other one.

7. Conclusion

In the global context of logistics and supply chain management, we are interested in the manner to model the SC. A Complex Adaptive Model, CAS, approach is then well studied for modelling supply chain systems considering the structural and behavioural dynamics. In a CAS, the interactions of the agent population and the environment evolution contribute to the emergence of a global behaviour.

This chapter presents an approach to study warehouse of chemical substances involving human actors. We have modelled the activities and the actors to implement a Multi-Agent System, MAS from which we want to reproduce segregation violation during the goods movements. Then, the warehouse becomes a CAS where agents accomplish their goals (typically handling operations) and whose mutual interactions are susceptible to violate segregations.

We propose a dynamic graph to describe the agents movements in the warehouse. Then, vertices represent agents when they are in activities and removed once they move back t their base. Edges are defined between vertices and are weighted by the distance between agents. The study of this graph by topological measures such as the average distance, the diameter and the mean degree show that agents are effectively close each other during their handling operations. We deduce a dynamic risk management to maintain segregation even when chemical substances are handled by agents. Thus, when the distance between incompatible goods is insufficient, a study of the two involved agents node betweenness determine what agent is redirected into another corridor. By this way the crossing and the mixing of incompatible goods is mitigated.

8. References

Albert, S., Jeong, H. & Barabasi, A.-L. (1999). The diameter of the world-wide web, *Nature* Vol. 401(No. 4): 130–131.

Bellifemine, F., Caire, G., Trucco, T. & Rimassa, G. (2007). Jade programmer's guide.

Bellifemine, F. L., Caire, G. & Greenwood, D. (2007). *Developing Multi-Agent Systems with JADE*, Wiley, MA, USA.

Broder, A., Kumar, R., Maghoul, F., Raghavan, P., Rajagopalan, S., Stata, R., Tomkins, A. & Wiener, J. (2000). Graph structure in the web, *Computer Networks* Vol. 33(No. 1): 309–320.

Burmeister, B., Haddadi, A. & Matylis, G. (1997). Application of multi-agent systems in traffic and transportation, *IEE Proceedings of Software Engineering*, IEEE Computer Society, London, pp. 51–60.

Casterfranchi, C. (1995). Guarantees for autonomy in cognitive agent architecture, *in* M. Wooldridge & N. R. Jennings (eds), *Lecture Notes in Computer Science: Intelligent Agents*, Springer-Verlag, Heidelberg, Germany, pp. 56–70.

Colla, V. & Nastasi, G. (2010). Modelling and simulation of an automated warehouse for the comparison of storage strategies, *in* G. Romero Rey & L. Martinez Muneta (eds), *Modelling Simulation and Optimization*, InTech, Rijeka, Croatia, pp. 471–486.

Conte, R. (1999). Social intelligence among autonomous agents, *Computational & Mathematical Organization Theory* Vol. 5(No. 3): 203–228.

Doran, J. (n.d.). Can agent-based modelling really be useful?, *in* N. J. Saam & B. Schmidt (eds), *Cooperative Agents: Applications in the Social Sciences*, Springer, pp. 57–81.

Epstein, J. & Axtell, R. (1996). *Growing Artificial Societies, Social Science from the bottom up*, Cambridge MA MIT press, MA, USA.

Erdos, P. & Rényi, A. (1959). On random graphs I, *Publicationes Mathematicae* Vol. 6(No. 1): 290–297.

Ferber, J. (1999). *Multi-Agent System: An Introduction to Distributed Artificial Intelligence*, Harlow: Addison Wesley Longman, MA, USA.

Flexsim Simulation Software (n.d.).
   URL: *http://http://www.flexsim.com/*

Forrester, J. W. (1961). *Industrial Dynamics*, Cambridge MA MIT press, MA, USA.

Gilbert, N. (2005). When does social simulation need cognitive models, *in* R. Sun (ed.), *Cognition and multi-agent interaction: From cognitive modeling to social simulation*, Cambridge University Press, Cambridge, UK, pp. 428–432.

Harary, F. & Gupta, G. (1997). Dynamic graph models, *Mathematical and Computer Modelling* Vol. 25(No. 7): 79–87.

Holland, J. H. (1996). *Hidden order: how adaptation builds complexity*, Reading, MA, Addison Wesley, MA, USA.

Jennings, N. R., Sycara, K. & Wooldridge, M. (1998). A roadmap of agent research and development, *Autonomous Agents and Multi-Agent Systems* Vol. 1(No. 1): 7–38.

Jeong, H., Tombor, B., Albert, R., Oltvai, Z. N. & Barabasi, A.-L. (2000). The large-scale organization of metabolic networks, *Nature* Vol. 6804(No. 407): 651–654.

Julka, N., Karimi, I. & Srinivasan, R. (2002). Agent-based refinery supply chain management, *Computer Aided Chemical Engineering* Vol. 26(No. 12): 1771–1781.

Kwon, O., Im, G. P. & Lee, K. C. (2006). Mace-scm: A multi-agent and case-based reasoning collaboration mechanism for supply chain management under supply and demand uncertainties, *Expert Systems with Applications* Vol. 33(No. 3): 690–705.

Lehman, J. F., Laird, J. & Rosenbloom, P. (2006). A gentel introduction to soar, an architecture for human cognition.

Pollack, M. E. & Ringuette, M. (1990). Introducing the tileworld: Experimentally evaluating agent architectures, *Proceedings the Eighth National Conference on Artificial Intelligence*, AAAI Press, Boston, MA, USA, pp. 183–189.

Reynolds, C. W. (1971). Flocks, herds and schools: A distributed behavioural model, *Computer Graphics* Vol. 21(No. 4): 25–34.

Schelling, T. (1971). Dynamic models of segregation, *Journal of Mathematical Sociology* Vol. 1(No. 2): 143–186.

Sloman, A. & Logan, B. (1999). Building cognitively rich agents using the sim agent toolkit, *Communications of the Association of Computing Machinery* Vol. 42(No. 3): 71–77.

Sun, R. (2001). Cognitive science meets multi-agent systems: A prolegomenon, *Philosophical Psychology* Vol. 14(No. 1): 5–28.

Swaminathan, J. M., Smith, S. F. & Sadeh, N. M. (1997). Modeling supply chain dynamics: A multi-agent approach, *Decision Sciences* Vol. 29(No. 3): 607–632.

Taatgen, N., Lebiere, C. & Anderson, J. (n.d.). Modeling paradigms in act-r, *in* R. Sun (ed.), *Cognition and Multi-Agent Interaction: From Cognitive Modelling to Social Simulation*, Cambridge University Press, pp. 29–52.

Wasserman, S. & Faust, K. (1994). *Social network analysis: methods and applications*, Cambridge University Press, Cambridge, UK.

Watts, D. J. & Strogatz, S. H. (1998). Collective dynamics of 'small-world' networks, *Nature* Vol. 393(No. 6684): 440–442.

Weiming, S., Qi, H., Hyun, J. Y. & Douglas, H. N. (2000). Applications of agent-based systems in intelligent manufacturing: An updated review, *Advanced Engineering Informatics* Vol. 20(No. 4): 415–431.

Wray, R. E. & Jones, R. M. (n.d.). An introduction to soar as an agent architecture, *in* R. Sun (ed.), *Cognition and Multi-Agent Interaction: From Cognitive Modelling to Social Simulation*, Cambridge University Press, pp. 53–78.

Permissions

The contributors of this book come from diverse backgrounds, making this book a truly international effort. This book will bring forth new frontiers with its revolutionizing research information and detailed analysis of the nascent developments around the world.

We would like to thank Yagang Zhang, for lending his expertise to make the book truly unique. He has played a crucial role in the development of this book. Without his invaluable contribution this book wouldn't have been possible. He has made vital efforts to compile up to date information on the varied aspects of this subject to make this book a valuable addition to the collection of many professionals and students.

This book was conceptualized with the vision of imparting up-to-date information and advanced data in this field. To ensure the same, a matchless editorial board was set up. Every individual on the board went through rigorous rounds of assessment to prove their worth. After which they invested a large part of their time researching and compiling the most relevant data for our readers. Conferences and sessions were held from time to time between the editorial board and the contributing authors to present the data in the most comprehensible form. The editorial team has worked tirelessly to provide valuable and valid information to help people across the globe.

Every chapter published in this book has been scrutinized by our experts. Their significance has been extensively debated. The topics covered herein carry significant findings which will fuel the growth of the discipline. They may even be implemented as practical applications or may be referred to as a beginning point for another development. Chapters in this book were first published by InTech; hereby published with permission under the Creative Commons Attribution License or equivalent.

The editorial board has been involved in producing this book since its inception. They have spent rigorous hours researching and exploring the diverse topics which have resulted in the successful publishing of this book. They have passed on their knowledge of decades through this book. To expedite this challenging task, the publisher supported the team at every step. A small team of assistant editors was also appointed to further simplify the editing procedure and attain best results for the readers.

Our editorial team has been hand-picked from every corner of the world. Their multi-ethnicity adds dynamic inputs to the discussions which result in innovative outcomes. These outcomes are then further discussed with the researchers and contributors who give their valuable feedback and opinion regarding the same. The feedback is then collaborated with the researches and they are edited in a comprehensive manner to aid the understanding of the subject.

Apart from the editorial board, the designing team has also invested a significant amount of their time in understanding the subject and creating the most relevant covers. They scrutinized every image to scout for the most suitable representation of the subject and create an appropriate cover for the book.

The publishing team has been involved in this book since its early stages. They were actively engaged in every process, be it collecting the data, connecting with the contributors or procuring relevant information. The team has been an ardent support to the editorial, designing and production team. Their endless efforts to recruit the best for this project, has resulted in the accomplishment of this book. They are a veteran in the field of academics and their pool of knowledge is as vast as their experience in printing. Their expertise and guidance has proved useful at every step. Their uncompromising quality standards have made this book an exceptional effort. Their encouragement from time to time has been an inspiration for everyone.

The publisher and the editorial board hope that this book will prove to be a valuable piece of knowledge for researchers, students, practitioners and scholars across the globe.

List of Contributors

T. D. Sudhakar
St Joseph's College of Engineering, Chennai, India

K.J. Abraham
Programa de Pós Graduação em Genética, Faculdade de Medicina de Ribeirão Preto, Universidade de São Paulo, Ribeirão Preto SP, Brazil

Rohan Fernando
Dept. of Animal Science, Iowa State University, Ames IA, USA

Marc J. Richard
Department of Mechanical Engineering, Laval University, Québec (Québec), Canada

Weihua Li
School of Mech & Auto Eng, Souch China University of Technology, China
NSFI/UCRC Center for Intelligent Maintenance System, University of Cincinnati, USA

Wen Liu
School of Mech. & Auto Eng, Souch China University of Technology, China

Yan Chen and Jay Lee
NSFI/UCRC Center for Intelligent Maintenance System, University of Cincinnati, USA

Giovanni Scardoni and Carlo Laudanna
Center for BioMedical Computing (CBMC), University of Verona, Italy

Andrey Vavilin and Kang-Hyun Jo
University of Ulsan, Korea

Daryoush Habibi and Quoc Viet Phung
Edith Cowan University, Australia

Marco Antonio Garcia Carvalho and André Luis Costa
School of Technology, University of Campinas, Brazil

Ewa Grandys
Academy of Management in Łódź, Poland

Paulo Sérgio Sausen, Airam Sausen and Mauricio de Campos
Master's Program in Mathematical Modeling, Regional University of Northwestern Rio, Grande do Sul State (UNIJUI), Brazil

Pierre Mazzega
UMR GET, IRD, CNRS, Université de Toulouse III, France
International Joint Laboratory OCE, IRD – Universidade de Brasilia, Brazil

Romain Boulet and Thérèse Libourel
UMR ESPACE-DEV, IRD, Université de Montpellier II, France

Omar Gaci
ISEL, Le Havre University, France

Hervé Mathieu
LITIS - ISEL, Le Havre University, France